IEE RADAR, SONAR AND NAVIGATION SERIES 19

Series Editors: Dr N. Stewart
Professor H. Griffiths

Radar Imaging and Holography

Other volumes in this series:

Volume 1	**Optimised radar processors** A. Farina (Editor)
Volume 2	**Radar development to 1945** R. W. Burns (Editor)
Volume 3	**Weibull radar clutter** M. Sekine and Y. H. Mao
Volume 4	**Advanced radar techniques and systems** G. Galati (Editor)
Volume 7	**Ultrawideband radar measurements** L. Yu. Astanin and A. A. Kostylev
Volume 8	**Aviation weather surveillance systems: advanced radar and surface sensors for flight safety and air traffic management** P. Mahapatra
Volume 10	**Radar techniques using array antennas** W. D. Wirth
Volume 11	**Air and spaceborne radar systems** P. Lacomme, J. C. Marchais, J. P. Hardange and E. Normant
Volume 12	**Principles of space-time adaptive processing** R. Klemm
Volume 13	**Introduction to RF Stealth** D. Lynch
Volume 14	**Applications of space-time adaptive processing** R. Klemm (Editor)
Volume 15	**Ground penetrating radar, 2nd edition** D. J. Daniels (Editor)
Volume 16	**Target detection by marine radar** J. Briggs
Volume 17	**Strapdown inertial navigation technology** D. H. Titterton and J. L. Weston

Radar Imaging and Holography

Alexander Ya. Pasmurov and
Julius S. Zinoviev

The Institution of Electrical Engineers

Published by:

The Institution of Electrical Engineers,
Michael Faraday House,
Six Hills Way, Stevenage,
Herts. SG1 2AY, United Kingdom

© 2005: The Institution of Electrical Engineers

This publication is copyright under the Berne Convention and the Universal Copyright Convention. All rights reserved. Apart from any fair dealing for the purposes of research or private study, or criticism or review, as permitted under the Copyright, Designs and Patents Act, 1988, this publication may be reproduced, stored or transmitted, in any forms or by any means, only with the prior permission in writing of the publishers, or in the case of reprographic reproduction in accordance with the terms of licences issued by the Copyright Licensing Agency. Inquiries concerning reproduction outside those terms should be sent to the IEE at the address above.

While the authors and the publishers believe that the information and guidance given in this work are correct, all parties must rely upon their own skill and judgment when making use of them. Neither the authors nor the publishers assume any liability to anyone for any loss or damage caused by any error or omission in the work, whether such error or omission is the result of negligence or any other cause. Any and all such liability is disclaimed.

The moral right of the authors to be identified as authors of this work have been asserted by them in accordance with the Copyright, Designs and Patents Act 1988.

British Library Cataloguing in Publication Data

Pasmurov, Alexander Ya.
   Radar imaging and holography
    1. Radar 2. Imaging systems 3. Radar targets 4. Holography
    I. Title II. Zinoviev, Julius S. III. Institution of Electrical Engineers
    621.3'848

ISBN-10: 0 86341 502 4
ISBN-13: 978-086341-502-9

Typeset in India by Newgen Imaging Systems (P) Ltd., Chennai, India
Printed in the UK by MPG Books Limited, Bodmin, Cornwall

Contents

List of figures   ix

List of tables   xvii

Introduction   1

1 Basic concepts of radar imaging   7
   1.1 Optical definitions   7
   1.2 Holographic concepts   10
   1.3 The principles of computerised tomography   14
   1.4 The principles of microwave imaging   20

2 Methods of radar imaging   27
   2.1 Target models   27
   2.2 Basic principles of aperture synthesis   31
   2.3 Methods of signal processing in imaging radar   33
       2.3.1 SAR signal processing and holographic radar for earth surveys   33
       2.3.2 ISAR signal processing   34
   2.4 Coherent radar holographic and tomographic processing   36
       2.4.1 The holographic approach   36
       2.4.2 Tomographic processing in 2D viewing geometry   41

3 Quasi-holographic and holographic radar imaging of point targets on the earth surface   49
   3.1 Side-looking SAR as a quasi-holographic radar   49
       3.1.1 The principles of hologram recording   50
       3.1.2 Image reconstruction from a microwave hologram   53
       3.1.3 Effects of carrier track instabilities and object's motion on image quality   57

	3.2	Front-looking holographic radar	60
		3.2.1 The principles of hologram recording	60
		3.2.2 Image reconstruction and scaling relations	62
		3.2.3 The focal depth	67
	3.3	A tomographic approach to spotlight SAR	70
		3.3.1 Tomographic registration of the earth area projection	70
		3.3.2 Tomographic algorithms for image reconstruction	72

4 Imaging radars and partially coherent targets — 79
    4.1 Imaging of extended targets — 80
    4.2 Mapping of rough sea surface — 82
    4.3 A mathematical model of imaging of partially coherent extended targets — 85
    4.4 Statistical characteristics of partially coherent target images — 87
        4.4.1 Statistical image characteristics for zero incoherent signal integration — 88
        4.4.2 Statistical image characteristics for incoherent signal integration — 90
    4.5 Viewing of low contrast partially coherent targets — 94

5 Radar systems for rotating target imaging (a holographic approach) — 101
    5.1 Inverse synthesis of 1D microwave Fourier holograms — 101
    5.2 Complex 1D microwave Fourier holograms — 110
    5.3 Simulation of microwave Fourier holograms — 112

6 Radar systems for rotating target imaging (a tomographic approach) — 117
    6.1 Processing in frequency and space domains — 117
    6.2 Processing in 3D viewing geometry: 2D and 3D imaging — 119
        6.2.1 The conditions for hologram recording — 120
        6.2.2 Preprocessing of radar data — 124
    6.3 Hologram processing by coherent summation of partial components — 126
    6.4 Processing algorithms for holograms of complex geometry — 130
        6.4.1 2D viewing geometry — 131
        6.4.2 3D viewing geometry — 141

7 Imaging of targets moving in a straight line — 147
    7.1 The effect of partial signal coherence on the cross range resolution — 148
    7.2 Modelling of path instabilities of an aerodynamic target — 151
    7.3 Modelling of radar imaging for partially coherent signals — 152

8	**Phase errors and improvement of image quality**		**157**
	8.1	Phase errors due to tropospheric and ionospheric turbulence	157
		8.1.1 The refractive index distribution in the troposphere	157
		8.1.2 The distribution of electron density fluctuations in the ionosphere	166
	8.2	A model of phase errors in a turbulent troposphere	167
	8.3	A model of phase errors in a turbulent ionosphere	172
	8.4	Evaluation of image quality	173
		8.4.1 Potential SAR characteristics	173
		8.4.2 Radar characteristics determined from images	175
		8.4.3 Integral evaluation of image quality	177
	8.5	Speckle noise and its suppression	181
		8.5.1 Structure and statistical characteristics of speckle	182
		8.5.2 Speckle suppression	184
9	**Radar imaging application**		**191**
	9.1	The earth remote sensing	191
		9.1.1 Satellite SARs	191
		9.1.2 SAR sea ice monitoring in the Arctic	195
		9.1.3 SAR imaging of mesoscale ocean phenomena	204
	9.2	The application of inverse aperture synthesis for radar imaging	215
	9.3	Measurement of target characteristics	217
	9.4	Target recognition	222

References   **231**

List of abbreviations   **241**

Index   **243**

List of figures

Chapter 1

Figure 1.1	The process of imaging by a thin lens	8
Figure 1.2	A schematic illustration of the focal depth of an optical image: (a) image of point M lying in the optical axis; (b) image of point A; (c) image of point B and (d) image of points A and B in the planes M_1, M_2 and M_3	9
Figure 1.3	The process of optical hologram recording: 1 – reference wave; 2 – object; 3 – photoplate and 4 – object's wave	11
Figure 1.4	Image reconstruction from a hologram: 1 – virtual image; 2 – real image; 3 – zero diffraction order and 4 – hologram	13
Figure 1.5	Viewing geometry in computerised tomography (from Reference 15): Γ_m – circumference for measurements; Γ_c – circumference with the centre at point O enveloping a cross section; p – arbitrary point in the circle with the polar coordinates ρ and Φ; A, C and D – wide beam transmitters; B, C' and D' – receivers; γ–γ, δ–δ – parallel elliptic arcs defining the resolving power of transmitter–receiver pair (CC' and DD')	15
Figure 1.6	A scheme of X-ray tomographic experiment using a collimated beam: 1 – X-rays; 2 – projection angle; 3 – registration line; 4 – projection axis and 5 – integration line	17
Figure 1.7	The geometrical arrangement of the $G(x,y)$ pixels in the Fourier region of a polar grid. The parameters ϑ_{max} and ϑ_{min} are the variation range of the projection angles. The shaded region is the SAR recording area	18
Figure 1.8	Synthesis of a radar aperture pattern: (a) real antenna array and (b) synthesised antenna array	22

x   List of figures

Chapter 2

Figure 2.1	Viewing geometry for a rotating cylinder: 1, 2, 3 – scattering centres (scatterers)	30
Figure 2.2	Schematic illustrations of aperture synthesis techniques: (a) direct synthesis implemented in SAR, (b) inverse synthesis for a target moving in a straight line and (c) inverse synthesis for a rotating target	32
Figure 2.3	The holographic approach to signal recording and processing in SAR: 1 – recording of a 1D Fraunhofer or Fresnel diffraction pattern of target field in the form of a transparency (azimuthal recording of a 1D microwave hologram), 2 – 1D Fourier or Fresnel transformation, 3 – display	34
Figure 2.4	Synthesis of a microwave hologram: (a) quadratic hologram recorded at a high frequency, (b) quadratic hologram recorded at an intermediate frequency, (c) multiplicative hologram recorded at a high frequency, (d) multiplicative hologram recorded at an intermediate frequency, (e) quadrature holograms, (f) phase-only hologram	37
Figure 2.5	A block diagram of a microwave holographic receiver: 1 – reference field, 2 – reference signal $\cos(\omega_0 t + \varphi_0)$, 3 – input signal $A \cos(\omega_0 t + \varphi_0 - \varphi)$, 4 – signal $\sin(\omega_0 t + \varphi_0)$ and 5 – mixer	39
Figure 2.6	Illustration for the calculation of the phase variation of a reference wave	39
Figure 2.7	The coordinates used in target viewing	42
Figure 2.8	2D data acquisition design in the tomographic approach	45
Figure 2.9	The space frequency spectrum recorded by a coherent (microwave holographic) system. The projection slices are shifted by the value f_{po} from the coordinate origin	46
Figure 2.10	The space frequency spectrum recorded by an incoherent (tomographic) system	47

Chapter 3

Figure 3.1	A scheme illustrating the focusing properties of a Fresnel zone plate: 1 – collimated coherent light, 2 – Fresnel zone plate, 3 – virtual image, 4 – real image and 5 – zeroth-order diffraction	50
Figure 3.2	The basic geometrical relations in SAR	51
Figure 3.3	An equivalent scheme of 1D microwave hologram recording by SAR	51
Figure 3.4	The viewing field of a holographic radar	60
Figure 3.5	A schematic diagram of a front-looking holographic radar	61
Figure 3.6	The resolution of a front-looking holographic radar along the x-axis as a function of the angle φ	61

Figure 3.7	The resolution of a front-looking holographic radar along the z-axis as a function of the angle φ	62
Figure 3.8	Generalised schemes of hologram recording (a) and reconstruction (b)	63
Figure 3.9	Recording (a) and reconstruction (b) of a two-point object for finding longitudinal magnifications: 1, 2 – point objects, 3 – reference wave source and 4 – reconstructing wave source	67
Figure 3.10	The focal depth of a microwave image: 1 – reconstructing wave source, 2 – real image of a point object and 3 – microwave hologram	68
Figure 3.11	The basic geometrical relations for a spot-light SAR	70

Chapter 4

Figure 4.1	The geometrical relations in a SAR	85
Figure 4.2	A generalised block diagram of a SAR	85
Figure 4.3	The variation of the parameter Q with the synthesis range L_s at $\lambda = 3$ cm, $\Theta = 0.02$ and various values of R	90
Figure 4.4	The dependence of the spatial correlation range of the image on normalised L_s for multi-ray processing (solid lines) at various degrees of incoherent integration D_e and for averaging of the resolution elements (dashed lines) at various G_e : $\lambda = 3$ cm, $R = 10$ km; 1, 5–0 (curves overlap); 2, 6–0.25$(\lambda R/2)^{1/2}$; 3, 7–$(\lambda R/2)^{1/2}$; 4, 8–2.25$(\lambda R/2)^{1/2}$	91
Figure 4.5	The variation of the parameter Q_h with the number of integrated signals N_i at various values of K_a	93
Figure 4.6	The variation of the parameter Q_e with the synthesis range L_s at various signal correlation times τ_c	97
Figure 4.7	The parameter Q as a function of the synthesis range L_s at various signal correlation times τ_c	98

Chapter 5

Figure 5.1	A schematic diagram of direct bistatic radar synthesis of a microwave hologram along arc L of a circle of radius R_0: 1 – transmitter, 2 – receiver	102
Figure 5.2	A schematic diagram of inverse synthesis of a microwave hologram by a unistatic radar located at point C	103
Figure 5.3	The geometry of data acquisition for the synthesis of a 1D microwave Fourier hologram of a rotating object	103
Figure 5.4	Optical reconstruction of 1D microwave images from a quadrature Fourier hologram: (a) flat transparency, (b) spherical transparency	106
Figure 5.5	The dependence of microwave image resolution on the normalised aperture angle of the hologram	109

xii   List of figures

Figure 5.6   Microwave images reconstructed from Fourier holograms: (a) quadrature hologram, (b) complex hologram with carrier frequency, (c) complex hologram without carrier frequency and (d,e,f) the variation of the reconstructed image with the hologram angle ψ_s (complex hologram without carrier frequency)   113

Figure 5.7   The algorithm of digital processing of 1D microwave complex Fourier holograms   114

Figure 5.8   A microwave image of a point object, reconstructed digitally from a complex Fourier hologram as a function of the object's aspects $\Psi_0 (\Psi_s = \pi/6)$: (a) $\Psi_0 = \pi/12$, (b) $\Psi_0 = 5\pi/2$ and (c) $\Psi_0 = 3\pi/4$.   115

Chapter 6

Figure 6.1   The aspect variation relative to the line of sight of a ground radar as a function of the viewing time for a satellite at the culmination altitudes of 31°, 66° and 88°: (a) aspect α and (b) aspect β 121

Figure 6.2   Geometrical relations for 3D microwave hologram recording: (a) data acquisition geometry; a–b, trajectory projection onto a unit surface relative to the radar motion and (b) hologram recording geometry   123

Figure 6.3   The sequence of operations in radar data processing during imaging   125

Figure 6.4   Subdivision of a 3D microwave hologram into partial holograms: (a) 1D partial (radial and transversal), (b) 2D partial (radial and transversal) and (c) 3D partial holograms   128

Figure 6.5   Subdivision of a 3D surface hologram into partial holograms: (a) radial, (b) 1D partial transversal and (c) 2D partial   129

Figure 6.6   Coherent summation of partial hologram. A 2D narrowband microwave hologram: (a) highlighting of partial holograms and (b) formation of an integral image   132

Figure 6.7   Coherent summation of partial hologram. A 2D wideband microwave hologram: (a) highlighting of partial holograms, (b) formation of an integral image   133

Figure 6.8   The computational complexity of the coherent summation algorithms as a function of the target dimension for a narrowband microwave hologram: (a) transverse partial images, (b) hologram samples   137

Figure 6.9   The relative computational complexity of coherent summation algorithms as a function of the target dimension for a narrowband microwave hologram: (a) transverse partial images/CCA, (b) hologram samples/CCA   140

Figure 6.10	The relative computational complexity of coherent summation algorithms of hologram samples and transverse partial images versus the coefficient μ in the case of a wideband hologram	141
Figure 6.11	The relative computational complexity of coherent summation algorithms for radial and transverse partial images versus the coefficient μ in the case of a wideband hologram	142
Figure 6.12	The transformation of the partial coordinate frame in the processing of a 3D hologram by coherent summation of transverse partial images	143

Chapter 7

Figure 7.1	Characteristics of an imaging device in the case of partially coherent echo signals: (a) potential resolving power at $C_2 = 1$, (b) performance criterion ($1 - d_c = 6.98$ m, $2 - d_c = 3.49$ m and $3 - d_c = 0$)	151
Figure 7.2	Typical errors in the impulse response of an imaging device along the s-axis: (a) response shift, (b) response broadening, (c) increased amplitude of the response side lobes and (d) combined effect of the above factors	153
Figure 7.3	The resolving power of an imaging device in the presence of range instabilities versus the synthesis time T_s and the method of resolution step measurement: (a) $-\sigma_p = 0.04$ m; 1 and 1′ (2 and 2′) – first (second) way of resolution step measurement; 1 and 2 – $T_c = 1.5$ s, 1′ and 2′ – $T_c = 3$ s; (b) $-\sigma_p = 0.05$ m, 1 and 1′ (2 and 2′) – first (second) way of resolution step measurement; 1 and 2 – $T_c = 1.5$ s, 1′ and 2′ – $T_c = 3$ s	154
Figure 7.4	The resolving power of an imaging system in the presence of velocity instabilities versus the synthesis time T_s and the method of resolution step measurement: (a) $\sigma_{x',y'} = 0.1$ m/s (other details as in Fig. 7.3), (b) $\sigma_{x',y'} = 0.2$ m/s (other details as in Fig. 7.3)	155
Figure 7.5	Evaluation of the performance of a processing device in the case of partially coherent signals versus the synthesis time T_s and the space step of path instability correlation d_c: $1-d_c = 6.98$m, $2-d_c = 3.49$ m	155

Chapter 8

Figure 8.1	The normalised refractive index spectrum $\Phi_n(\chi)/C_n^2$ as a function of the wave number χ in various models: 1 – Tatarsky's model-I, 2 – Tatarsky's model-II, 3 – Carman's model and 4 – modified Carman's model	160
Figure 8.2	The profile of the structure constant C_n^2 versus the altitude for April at the SAR wavelength of 3.12 cm	164
Figure 8.3	The profile of the structure constant C_n^2 versus the altitude for November at the SAR wavelength of 3.12 cm	165

xiv List of figures

Figure 8.4	A geometrical construction for a spaceborne SAR tracking a point object A through a turbulent atmospheric stratum of thickness h_t	168
Figure 8.5	A schematic test ground with corner reflectors for investigation of SAR performance	176
Figure 8.6	A 1D SAR image of two corner reflectors	177
Figure 8.7	A histogram of the noise distribution in a SAR receiver	178
Figure 8.8	The grey-level (half-tone) resolution versus the number of incoherently integrated frames N	179
Figure 8.9	The dependence of the image interpretability on the resolution versus linear resolution $p_a = p_r = p$	180
Figure 8.10	The dependence of the half-tone resolution on the number of incoherent integrations over the total real antenna pattern	181

Chapter 9

Figure 9.1	The mean monthly convoy speed in the NSR changes from V_0 (without satellite data) to V_1 (SAR images used by the icebreaker's crew to select the route in sea ice). The mean ice thickness (h_i) is shown as a function of the season. (N. Babich, personal communications)	198
Figure 9.2	(a) Photo of grease ice and (b) a characteristic dark SAR signature of grease ice. ©European Space Agency	199
Figure 9.3	Photo of typical nilas with finger-rafting	200
Figure 9.4	A RADARSAT ScanSAR Wide image of 25 April 1998, covering an area of 500 km × 500 km around the northern Novaya Zemlya. A geographical grid and the coastline are superimposed on the image. ©Canadian Space Agency	201
Figure 9.5	A RADARSAT ScanSAR Wide image of 3 March 1998, covering the boundary between old and first-year sea ice in the area to north Alaska. ©Canadian Space Agency	202
Figure 9.6	(a) Photo of a typical pancake ice edge and (b) a characteristic ERS SAR signature of pancake ice. A mixed bright and dark backscatter signature is typical for pancake and grease ice found at the ice edge. ©European Space Agency	203
Figure 9.7	A RADARSAT ScanSAR Wide image of 8 May 1998, covering the south-western Kara Sea. ©Canadian Space Agency	204
Figure 9.8	An ENVISAT ASAR image of 28 March 2003, covering the ice edge in the Barents Sea westward and southward of Svalbard. ©European Space Agency	205
Figure 9.9	An ERS-2 SAR image of 11 September 2001, covering the Red Army Strait in the Severnaya Zemlya Archipelago. ©European Space Agency	206

List of figures   xv

Figure 9.10	An ERS-2 SAR image (100 km × 100 km) taken on 24 June 2000 over the Black Sea (region to the East Crimea peninsula) and showing upwelling, natural films	208
Figure 9.11	SST retrieved from a NOAA AVHRR image on 24 June 2000.	209
Figure 9.12	A fragment of an ERS-2 SAR image (26 km × 22 km) taken on 30 September 1995 over the Northern Sea near the Norwegian coast and showing swell	210
Figure 9.13	An ERS-2 SAR image (100 km × 100 km) taken on 28 September 1995 over the Northern Sea and showing an oil spill, wind shadow, low wind and ocean fronts	211
Figure 9.14	An ERS-1 SAR image (100 km × 100 km) taken on 29 September 1995 over the Northern Sea showing rain cells	213
Figure 9.15	An ERS-2 SAR image (18 km × 32 km) taken on 30 September 1995 over the Northern Sea showing an internal wave and a ship wake	214
Figure 9.16	The scheme of the reconstruction algorithm	221
Figure 9.17	A typical 1D image of a perfectly conducting cylinder	222
Figure 9.18	The local scattering characteristics for a metallic cylinder (E-polarisation)	223
Figure 9.19	The local scattering characteristics for a metallic cylinder (H-polarisation)	224
Figure 9.20	A mathematical model of a radar recognition device	226

List of tables

Chapter 6
Table 6.1　The number of spectral components of a PH　136

Chapter 8
Table 8.1　The main characteristics of the synthetic aperture pattern　174

Chapter 9
Table 9.1　Technical parameters of SARs borne by the SEASAT and Shuttle　192
Table 9.2　Parameters of the Almaz-1 SAR　192
Table 9.3　The parameters of the ERS-1/2 satellites　193
Table 9.4　SAR imaging modes of the RADARSAT satellite　194
Table 9.5　The ENVISAT ASAR operation modes　194
Table 9.6　The LRIR characteristics　216
Table 9.7　The variants of the sign vectors　227
Table 9.8　The valid recognition probability (a Bayes classifier)　228
Table 9.9　The valid recognition probability (a classifier based on the method of potential functions)　228

Introduction

The analysis of the current state and tendencies in radar development shows that novel methods of target viewing are based on a detailed study of echo signals and their informative characteristics. These methods are aimed at obtaining complete data on a target, with emphasis on revealing new steady parameters for their recognition. One way of raising the efficiency of radar technology is to improve available methods of radio vision, or imaging. Radio vision systems provide a high resolution, considerably extending the scope of target detection and recognition. This field of radar science and technology is very promising, because it paves the way from the classical detection of a point target to the imaging of a whole object.

The physical mechanism underlying target viewing can be understood on a heuristic basis. An electromagnetic wave incident on a target induces an electric current on it, generating a scattered electromagnetic wave. In order to find the scattering properties of the target, we must visualise its elements making the greatest contribution to the wave scattering. This brings us to the concept of a radar image, which can be defined as a spatial distribution pattern of the target reflectivity. Therefore, an image must give a spatial quantitative description of this physical property of the target with a quality not less than that provided by conventional observational techniques.

Radio vision makes it possible to sense an object as a visual picture. This is very important because we get about 90 per cent of all information about the world through vision. Of course, a radar image differs from a common optical image. For instance, a surface rough to light waves will be specular to radio waves (microwaves), and images of many objects will look like bright spots, or glare. However, the representation of information transported by microwaves as visual images has become quite common. It took much time and effort to get a high angular resolution in the microwave frequency band because of the limited size of a real antenna. It was not until the 1950–1960s that a sufficiently high resolution was obtained by a side-looking radar with a large synthesised antenna aperture. The synthetic aperture method was then described in terms of the range-Doppler approach.

At about the same time, a new method of imaging in the visible spectrum emerged which was based on recording and reconstruction of the wave front and its phase, using a reference wave. A lens-free registration of the wave front (the holographic technique), followed by the image reconstruction, was first suggested by D. Gabor in

1948 and re-discovered by E. Leith and U. Upatnieks in 1963. The two researchers suggested a holographic method with a 'side reference beam' to eliminate the zeroth diffraction order. This principle was later used in a new, side-looking type of radar.

A specific feature of holographic imaging is that a hologram records an integral Fourier or Fresnel transform of the object's scattering function. The emergence of holography radically changed our conception of an object's image. Earlier, humans had dealt with images produced by recording the distribution of light intensity in a certain plane. But objects can generate a light field or another kind of electromagnetic field with all of its parameters modulated: the amplitude, phase, polarisation, etc. This discovery considerably extended the scope of spatial information that could be extracted about the object of interest.

It should be noted that holography brought about revolutionary changes only in optics, because it did not possess ways or means to save the recorded information about the phase structure of an optical field until then. But the application of holographic principles to the microwave frequency band proceeded easily, giving excellent results. This was due to the fact that radio engineering had employed methods of registration of the electromagnetic wave phase long before the emergence of holography. For many years, radar imaging developed independently of holography, although some workers (E.N. Leith, W.E. Kock, D.L. Mensa, B.D. Steinberg) did note that many intermediate steps in the recording and processing techniques for radar imaging were quite similar to those of holography and tomography. These researchers, however, only briefly reviewed the holographic principles just to point out the fundamental similarity and difference between optical and radar imaging, but they did not make a comprehensive analysis of this fact in the context of radiolocation.

E.N. Leith and A.L. Ingalls showed that the operation of a side-looking radar should be treated in terms of a holographic approach. Holograms recorded in the microwave frequency range were referred to as microwave holograms, and radar systems based on the holographic principle were called by E.N. Leith quasi-holographic. In fact, the work done in those years became the basis for designing a special type of radar to perform imaging. The research into radar imaging was developing quite intensively, and many scientists made their contributions to it: L.J. Cutrona, A. Kozma, D.A. Ausherman, G. Graf, I.L. Walker, W.M. Brown, D.L. Mensa, D.C. Manson, B.D. Steinberg, N.H. Farhat, V.C. Chen, D.R. Wehner and others.

These efforts were accompanied by the development of tomographic techniques for image reconstruction in medicine and physiology (X-ray imaging). Initially, tomography was treated as a way of reconstructing the spatial distribution of a certain physical characteristic of an object by making computational operations with data obtained during the probing of the object. This resulted in the emergence of reconstructive computerised tomography possessing powerful mathematical methods. Later, tomographic techniques were suggested capable of reconstructing a physical characteristic of an object by a mathematical processing of the field reflected by it.

Naturally, there have been suggestions to combine the available methods of radar imaging (e.g. the range-Doppler principles) with tomographic algorithms (D.L. Mensa, D.C. Manson). At present, the work on radar imaging goes on, combining the principles of microwave holography, range-Doppler methods of reflected field recording and tomographic image reconstruction.

In Russia, the theory of a side-looking radar has been developed by many workers: Yu.A. Melnik, N.I. Burenin, G.S. Kondratenkov, A.P. Reutov, Yu.A. Feoktistov, E.F. Tolstov, L.B. Neronsky and others. Much contribution to the theory of inverse aperture synthesis and tomographic image processing has been made by S.A. Popov, B.A. Rozanov, J.S. Zinoviev, A.Ya. Pasmurov, A.F. Kononov, A.A. Kuriksha, A. Manukyan and others.

This book presents systematised results on the application of direct and inverse aperture synthesis for radar imaging by holographic and tomographic techniques. The focus is on the research data obtained by the authors themselves. The book is primarily intended for engineers, designers and researchers, who are working in radar design and maintenance and are interested in the fundamental problem of extracting useful information from radar data.

The book consists of three parts: introductory Chapters 1 and 2, theoretical Chapters 3–8 and concluding Chapter 9.

The first two chapters will be useful to a reader who has but limited knowledge of optical holography, microwave holography and tomography. They cover the material available in the literature, but the information is presented in such a way that the reader will be able to better understand the chapters that follow. Besides, Chapter 1 treats the equation for an optical hologram in a non-trivial way to explain the speckle structure of a radar image. Chapter 2 explains the physical difference between coherent (microwave holographic) and incoherent (tomographic) imaging. The mathematical relations presented can be regarded as an extension of the classical theorem of a projection slice to coherent imaging. This allows application of the analytical methods of reconstructive computerised tomography for further development of coherent imaging theory.

Chapters 3–8 represent an attempt to treat the imaging radar operation in terms of holography, microwave holography and tomography, without resorting to the Doppler approach. Most of this material is the authors' results published during the past 30 years.

Chapter 3 discusses the holographic approach as applied to a side-looking radar. Its azimuthal channel is treated as a holographic system, in which the formation of a microwave hologram represents the recording of a field scattered from an artificial reference source, and the image reconstruction is described in terms of physical optics. We show that the use of a subcarrier frequency by turning the antenna beam away from the direction normal to the track velocity vector leads to a distorted image. The holographic approach can readily evaluate a permissible deviation of the carrier's pathway from a straight line and find various radar parameters, using conventional geometrical and optical methods. The holographic analysis of a front-looking radar on the basis of a generalised hologram geometry shows that the image is three-dimensional (3D); we describe the conditions for recording an undistorted image in the longitudinal and transversal directions. We also introduce the concept of a focal depth and explain the pseudoscopic character of an image. The application of tomographic principles to a spot-light radar is largely discussed using the results of D.S. Manson, who was the first to demonstrate their applicability to data processing.

Chapter 4 considers the radar aperture synthesis during the viewing of partially coherent and extended targets. The mathematical model of the aperture is also based on the holographic principle; the aperture is thought to be a filter with a frequency-contrast characteristic, which registers the space–time spectrum of a target. This approach is useful for the calculation of incoherent integration efficiency to smooth out low contrast details on an image.

In Chapter 5 we discuss microwave imaging of a rotating target, using 1D Fourier hologram theory and find the longitudinal and transverse scales of a reconstructed image, the target resolution and a criterion for an optimal processing of a Fourier microwave hologram. The resolution of a visual radar image is found to be consistent with the Abbe criterion for optical systems. One specificity is that it is necessary to introduce a space carrier frequency to separate two conjugate images and an image of the reference source. Here we have an analogy with synthetic aperture theory, with the exception that we employ the concept of a complex microwave Fourier hologram. It is shown that there is no zeroth diffraction order in digital reconstruction. We have formulated some requirements on methods and devices for synthesising this type of hologram. This method is easy and useful to implement in an anechoic chamber.

Chapter 6 focuses on tomographic processing of 2D and 3D microwave holograms of a rotating target in 3D viewing geometry with a non-equidistant arrangement of echo signal records in the registration of its aspect variation (for space objects). The suggested technique of image reconstruction is based on the processing of microwave holograms by coherent summation of partial holograms. These are classified into 1D, 2D, 2D radial, as well as narrowband and wideband partial holograms. This technique is feasible in any mode of target motion. The method of hologram synthesis combined with coherent computerised tomography represents a new processing technique which accounts for a large variation of real hologram geometries in 3D viewing. This advantage is inaccessible to other processing procedures yet.

Chapter 7 is concerned with methods of hologram processing for a target moving in a straight line and viewed by a ground radar processing partially coherent echo signals. The signal coherence is assumed to be perturbed by such factors as a turbulent medium, elastic vibrations of the target's body, vibrations of parts of the engines, etc. We suggest an approach to modelling the track instabilities of an aerodynamic target and present estimates of the radar resolving power in a real cross-section region.

Chapter 8 focuses on phase errors in radar imaging, evaluation of image quality and speckle noise.

Finally, possible applications of radar imaging are discussed in Chapter 9. The emphasis is on spaceborne synthetic aperture radars for surveying the earth surface. Some novel and original developments by researchers and designers at the Nansen Environmental and Remote Sensing Centre in Bergen (Norway) and at the Nansen International Environmental and Remote Sensing Centre in St Petersburg (Russia) are described. They have much experience in processing holograms from various SARs: Almaz-1 (Russia), RADARSAT (Canada), ERS-1/2 and ENVISAT ASAR (the European Space Agency). Of special interest to the reader might be the information about the use of microwave holography for classification of sea ice, navigation in the Arctics, a global monitoring of ocean phenomena and characteristics to be used for surveying gas and oil resources. We illustrate the use of the holographic methods in a coherent ground radar for 2D imaging of the Russian spacecraft Progress and for the study of local radar responses to objects of complex geometry in an anechoic chamber, aimed at target recognition.

To conclude, the methods and techniques described in this book are also applicable to many other research fields, including ultrasound and sonar, astronomy, geophysics, environmental sciences, resources surveys, non-destructive testing, aerospace defence and medical imaging, that have already started to utilise this rapidly developing technology. We hope that our book will also be used as an advanced textbook by postgraduate and graduate students in electrical engineering, physics and astronomy.

Acknowledgements

The idea to write a book about the application of holographic principles in radiolocation occurred to us at the end of the last century and was supported by the late Professor V.E. Dulevich. We are indebted to him for his encouragement and useful suggestions.

We express our gratitude to the staff members of the Nansen Centres (Bergens and St Petersburg), who provided us with valuable information about the practical application of a side-looking radar. We should like to thank V.Y. Aleksandrov, L.P. Bobylev, D.B. Akimov, O.M. Johannessen and S. Sandven for their help in the preparation of these materials.

Our deepest thanks also go to our colleagues E.F. Tolstov and A.S. Bogachev for their excellent description of the criteria for evaluation of radar images. This book is based on the results of our investigations that have taken a long period of time. We have collaborated with many specialists who helped to shape our conception of a coherent radar system. We thank them all, especially S.A. Popov, G.S. Kondratenkov, P.Ya. Ufimtzev, D.B. Kanareykin and Yu.A. Melnik, whose contribution was particularly valuable. We also thank our pupils V.R. Akhmetyanov, A.L. Ilyin and V.P. Likhachev for their assistance in the preparation of this book. We are also grateful to L.N. Smirnova, the translator of the book, for her immense help in producing the English version.

Chapter 1

Basic concepts of radar imaging

1.1 Optical definitions

At present, there is a certain class of microwave radars capable of imaging various types of extended targets. These are usually termed imaging radars. Before giving a definition of a 'microwave image', we should like to draw the reader's attention to two circumstances. First, a microwave image is always viewed by a radar operator in the visible range, while the imaging is performed in the microwave range. Second, this book considers radar imaging based on a combination of holographic and tomographic approaches. Therefore, we should first recall the basic concepts necessary for the description of imaging by conventional photographic and holographic devices in the visible spectral range.

Let us construct an image of an object (AB) formed by a thin lens (Fig. 1.1) [19]. The lens thickness can be neglected, and one can assume that the principal planes of the object AB and its image A'B' coincide and pass through the lens centre (line M'N'). The other designations are the focal lengths HF, HF', f, f' and the distances x, x' separating the object and its image from the respective focal points F and F'.

The straight line AA' connecting the vertices of the object and the image passes through the centre of the lens H. If we draw an auxiliary ray AF intercepting the principal plane at the point N and an auxiliary ray AM parallel to the optical axis at the point A', where the refracted rays MA' and NA' intercept, we can find the image of the point A. If we draw the normal A'B' from the point A' to the optical axis, we shall get the optical image of the object AB. The similarity conditions yield the governing equations for an optical image, or Newton's formulae:

$$\frac{y'}{y} = -\frac{f}{x} = -\frac{x'}{f'}, \qquad (1.1)$$

$$xx' = ff'. \qquad (1.2)$$

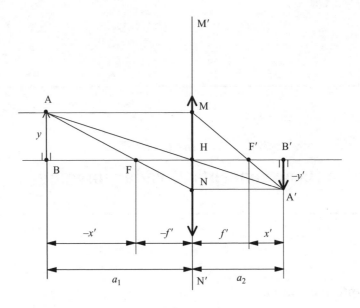

Figure 1.1 The process of imaging by a thin lens

The relation between the elements of an image and the corresponding elements of an object is known as a linear or transversal lens magnification V defined as

$$V = \frac{y'}{y}. \tag{1.3}$$

Since the lens is described by the equality $f = -f'$, Eq. (1.2) gives

$$xx' = -f^2. \tag{1.4}$$

Newton's formulae relate the distances of the object and the image to the respective focal points. However, it is sometimes more convenient to use their distances to the respective principal planes. Let us denote these distances as a_1 and a_2. Then using Fig. 1.1 and Eq. (1.2), we can get

$$\frac{1}{a_2} - \frac{1}{a_1} = \frac{1}{f_1}. \tag{1.5}$$

The linear magnification can be expressed through a_1 and a_2 as

$$V = \frac{a_2}{a_1}. \tag{1.6}$$

Consider now the concept of focal depth in the image space [80]. When constructing the image to be produced by a lens, we assumed that the image and the object were in planes normal to the optical axis. Suppose now that the object AB, say, a bulb filament, is inclined to the optical axis, as is shown in Fig. 1.2, while a photographic plate is in the plane M_1 normal to the optical axis of the objective lens. In order to

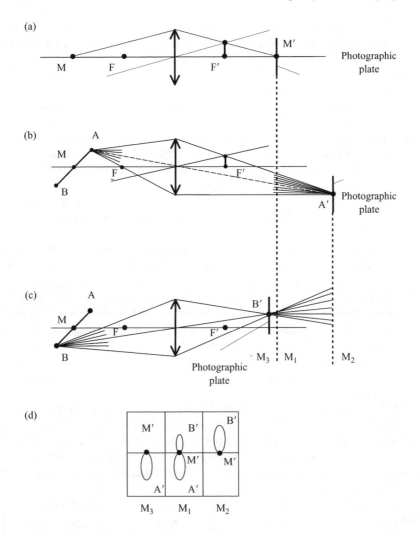

Figure 1.2 A schematic illustration of the focal depth of an optical image: (a) image of point M lying in the optical axis; (b) image of point A; (c) image of point B and (d) image of points A and B in the planes M_1, M_2 and M_3.

find the image on the photoplate, we shall construct rays of light going away from individual points of the object. The light beams going from the object AB to the objective lens and from the objective lens to the image are conic with the lens as the base and the points of the object and the image as the vertices. Imagine that the image of the point M of an object lying in the optical axis is on a photoplate in the plane M_1 (Fig. 1.2(a) and (d)). Then the beam of rays converging onto this image will have its vertex on the plate. The object's extremal points A and B will produce conic rays with the vertices in front of (B′) in Fig. 1.2(c) and behind the photoplate

(A′) (Fig. 1.2(b)). Thus, it is only the point M in the optical axis that will have its image as a bright point M′ in Fig. 1.2(a). The end points A and B of the line will look like light circles A′ and B′. The image of the line will look like M_1 in Fig.1.2(d). If the photoplate is shifted towards A′ (Fig. 1.2(b)) or B′ (Fig. 1.2(c)), we shall have different images M_2 or M_3 (Fig. 1.2(d)).

It follows from this representation that the image of a 3D object extended along the optical axis will have different focal depths on the plate at all the points in the image space. In practice, however, images of such objects have a good contrast. Therefore, the objective lens possesses a considerable focal depth. This parameter determines the longitudinal distance between two points of an object, and the sizes of their images do not exceed the eye's unit resolution. Therefore, the classical recording on a photoplate produces a 2D image, which cannot be transformed to a 3D image. The third dimension may be perceived only due to indirect phenomena such as the perspective.

Now let us describe the real and virtual optical images and see how the image of a point object M can be constructed with rays. The rays go away from the object in all directions. If one of the rays encounters a lens along its pathway, its trajectory will change. If the rays deflected by the lens intercept, when extended along the light propagation direction, a point image will be formed at the interception and can be recorded on a screen or a photoplate. This kind of image is known as real. However, when the rays are extended along the direction opposite to the light propagation direction, both the interception point and the image are said to be virtual. The images in Fig. 1.2 are real because they are formed by rays intercepting at their extension along the light propagation.

An optical image possesses orthoscopic and pseudoscopic properties. Suppose a 2D object has a surface relief, its image will be orthoscopic if it is not reversed longitudinally: the convex parts of the object look convex on the image. Using the above approach, we can show that the image formed by a thin lens is orthoscopic. If an image has a reverse relief, it is termed pseudoscopic; such images are produced by holographic cameras.

Thus, images produced by classical methods have the following typical characteristics.

- Imaging includes only the recording of incident light intensity, while its wave phase remains unrecorded. For this reason, this sort of image cannot be transformed to a 3D image.
- An image has a limited focal depth.
- An image produced by a thin lens is real and orthoscopic.

1.2 Holographic concepts

Holography is a lens-free way of recording images of 3D objects into 2D recording media [29]. This process includes two stages. The first stage is called hologram recording, during which the interference between the diffraction field from an object

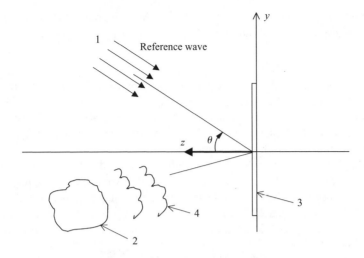

Figure 1.3 The process of optical hologram recording: 1 – reference wave; 2 – object; 3 – photoplate and 4 – object's wave

and a reference field is recorded on a photoplate or another photosensitive material. A necessary condition is that both fields should be coherent. In their original experiments, the pioneers of holography used mercury sources that were later replaced by lasers. The interference pattern registered on a photoplate was called a hologram. The second stage is that of image reconstruction including the illumination of the processed photoplate with a wave identical to the reference wave. Suppose, for simplicity, that the reference wave is plane (Fig. 1.3) and propagates at an angle θ to the z-axis (x, y, z are coordinates in the hologram plane). The object's wave is described by a complex function

$$u(x,y) = a(x,y)\exp(-j\varphi(x,y))$$

and the reference wave by the function

$$u_o(x,y) = a_o \exp(-j\omega_o x),$$

where $\omega_o = k\sin\theta$, θ is the wave incidence onto a photoplate located in the xOy plane, $k = 2\pi/\lambda_1$ is the wave number, and λ_1 is the wavelength of coherent light source.

The intensity of the interference pattern on the hologram is

$$\begin{aligned}I(x,y) &= [u_o(x,y) + u(x,y)]^2 = a_o^2 + a^2(x,y) \\ &\quad + \exp\{j[\varphi(x,y) - \omega_o x]\} + \exp\{-j[\varphi(x,y) - \omega_o x]\}a_o a(x,y) \\ &= a_o^2 + a^2(x,y) + 2a_o a(x,y)\cos[\varphi(x,y) - \omega_o x].\end{aligned} \quad (1.7)$$

In addition to the constant term $a_o^2 + a(x,y)$, the hologram function in Eq. (1.7) contains a harmonic term $2a_o a(x,y) \cos(\omega_o x)$ with the period

$$T = 2\pi/\omega_o = \lambda_1/\sin\theta. \tag{1.8}$$

The quantity ω_o which defines this period is known as the space carrier frequency (SCF) of a hologram. For example, for a He–Ne laser beam ($\lambda_1 = 0.6328\,\mu m$) incident onto a hologram at an angle of 30°, the SCF is $\omega_o = 900$ lines/mm. The minimum period of the SCF is $\theta = \pi/2$ and is equal to the wavelength λ_1. The a/a_o ratio is called the hologram modulation index.

It follows from Eq. (1.7) that the amplitude and phase distributions of the object's wave appear to be coded by the SCF amplitude and phase modulations, respectively. As a result, a hologram turns out to be the carrier of space frequency which contains spatial information, whereas a microwave is the carrier of angular frequency and contains temporal information. Phase-only holograms record only the phase variation rather than the amplitude.

The first stage of the holographic process is terminated by recording the quantity $I(x,y)$. A photoplate records a hologram. The transmittance of an exposed and processed photoplate is

$$T_n(x,y) = I^{-\gamma}, \tag{1.9}$$

where γ is the plate contrast coefficient. It is reasonable to take $\gamma = -2$ because the hologram then corresponds to a sine diffraction grating which does not form diffraction orders higher than the first one. So we have

$$t(x,y) = \sqrt{T_n(x,y)} = I(x,y).$$

During the reconstruction, a hologram is illuminated by the same reference wave as was used at the recording stage. The reconstruction occurs due to the light diffraction on the hologram (Fig. 1.4). Immediately behind the hologram, a wave field is induced with the following components:

$$U(x,y) = e^{-j\omega_o x} t(x,y) = e^{-j\omega_o x} I(x,y) = \exp(-j\omega_o x)[a_o^2 + a^2(x,y)]$$
$$+ a_o a(x,y) \exp\{-j\varphi(x,y)\} + \exp\{-j\omega_o x\} \exp\{j\varphi(x,y)\} a_o a(x,y). \tag{1.10}$$

With this, the second stage of the holographic process is terminated.

Three terms in Eq. (1.10) describe waves that form three different images (Fig. 1.4). The first wave preserves the direction of the reconstructing (plane) wave and represents the zero diffraction order, or light background. The second wave $a_o a(x,y) \exp[-j\varphi(x,y)]$ reproduces the object's wave to an accuracy of the amplitude factor a_o, providing a virtual image of the object observed behind the hologram. At an angle (-2θ) relative to the normal to the hologram, a complex conjugate wave propagates, producing a real image in front of the hologram. It can be shown (Chapter 3) that the virtual image is orthoscopic and the real image is pseudoscopic. Of importance is the fact that the virtual image is 3D.

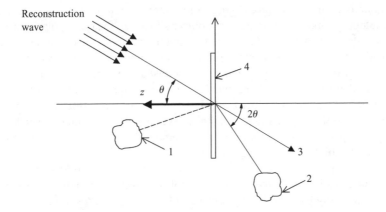

Figure 1.4 *Image reconstruction from a hologram: 1 – virtual image; 2 – real image; 3 – zero diffraction order and 4 – hologram*

Consider the basic properties of a holographic image, in particular, the hologram information structure. Suppose the object to be imaged is a discrete ensemble N of coherently radiating points with the coordinates r_q. The object's field on the hologram aperture can be described by a sum [108]

$$U_r = \sum_{q=1}^{N} a_q \exp(-jkr_q) = \sum_{q=1}^{N} \alpha_q \qquad (1.11)$$

and the respective intensity distribution by an expression

$$|U_r|^2 = \sum_{q=1}^{N} \sum_{p=1}^{N} \alpha_q \alpha_p^*, \qquad (1.12)$$

where the asterisk denotes a complex conjugate quantity.

The reference beam on the hologram aperture will be given as

$$U_o = a_o \exp(-jkr_o) = \alpha_o, \qquad (1.13)$$

where r_o are the reference beam coordinates. Then the intensity of the interference pattern can be written as

$$|U_r + U_o|^2 = a_o^2 + \alpha_o^* \sum_{q=1}^{N} \alpha_q + \alpha_o \sum_{p=1}^{N} \alpha_p^* + \sum_{q=1}^{N} \sum_{p=1}^{N} \alpha_q^* \alpha_p. \qquad (1.14)$$

The last term in Eq. (1.14) corresponds to Eq. (1.12) but usually it is not analysed completely. In holographic theory (Eq. (1.7)), one often restricts one's consideration to the second and third terms. Commonly, the information about the object is assumed to be distributed uniformly across the hologram aperture; in reality, however, a hologram is synthesised from a set of microholograms. So the aperture is split into a multiplicity of microapertures having various information significance. Such partial microholograms may correspond to the object's field with varying polarisation.

The hologram structure has three space-frequency levels. The first level is associated with the diffraction characteristics of individual radiating scatterers, more exactly, with their scattering patterns and the distance to the hologram plane. The second level is associated with the interference of overlapping fields of different radiating scatterers, a factor described by the last term in Eq. (1.14). Both levels determine the structure of the object's field. The third level is due to the interference between the object's speckle field and the reference beam field; this is a holographic structure possessing the highest space frequencies.

A scatterer reflects waves in all directions; therefore, every point of the hologram receives information about the object as a whole (the third information level). That is why we can easily explain the experiment with a hologram broken into pieces: any piece can reconstruct the whole image because it contains information from all the scatterers. If a piece is small, the image quality will be poor since some details are lost because of a poorer resolution. The result is a characteristic speckle pattern due to the greater effect of the second-order elements on the hologram.

Thus, holographic images have the following specific features.

- The holographic method of image recording registers the field phase in addition to the amplitude.
- Reconstructed images are 3D.
- Holographic images possess a considerable focal depth.
- Holographic images may be orthoscopic or pseudoscopic.

1.3 The principles of computerised tomography

Computerised tomography is generally defined as a method of reconstructing the true image (density distribution) of an object, using special computational procedures with data registered when the object is subjected to probing [15]. Generally, probing is an arbitrary physical phenomenon (radiation, wave propagation, etc.) used for the study of objects' structure, and density means the distribution of an arbitrary physical characteristic of the object to be reconstructed. The true image is an image, in which the reconstructed density at any point in space is ideally independent of the true densities beyond the point vicinity, or of the minimum object's volume resolvable by a measuring system. Since a probing wave interacts with the object and this interaction is 'integrated' along its passage through the object, it is clear why tomography is said to be a method of image reconstruction from integrated data such as beam sums, projections and so on. Therefore, computerised tomography is a way of producing 2D images of slices of 3D objects by means of digital processing of a multiplicity of 1D functions (projections) obtained at various vision angles. There are three important aspects of this technique. First, it is the problem of reconstructed image singularity, that is, the degree to which the object is describable by available data. Second, it is necessary to know whether the reconstruction process is resistant to errors and noise in the initial data. Finally, one must design an algorithm for image reconstruction.

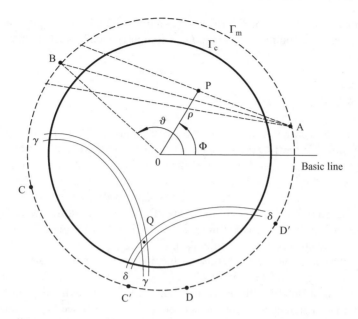

*Figure 1.5 Viewing geometry in computerised tomography (from Reference 15):
Γ_m – circumference for measurements; Γ_c – circumference with the
centre at point O enveloping a cross section; p – arbitrary point in the
circle with the polar coordinates ρ and Φ; A, C and D – wide beam
transmitters; B, C' and D' – receivers; γ–γ, δ–δ – parallel elliptic arcs
defining the resolving power of transmitter–receiver pair (CC' and DD')*

The principle of computerised tomography can be conveniently illustrated with a 2D case [15]. We shall first consider tomographic procedures for the reconstruction of density across a body's slice. Let us introduce a circumference Γ_c (Fig. 1.5) enveloping a body, more exactly, the cross section of a real 3D object. The inner Γ_c region can be termed the image space because it includes the object to be imaged. The medium outside this region is assumed to be free, which means that a probing wave interacts only with the object. If the probing sources are located outside the Γ_c region, the method is called remote-probing computerised tomography, as opposed to remote-sensing tomography when the sources are located within this region. The latter is, however, of no interest to radar imaging. The probing effects are commonly measured outside the Γ_c region. It is clear from information theory that measurements made along a certain circumference Γ_m (Fig. 1.5) embracing Γ_c will be quite sufficient.

Suppose a probing radiation transmitter is located in the circumference Γ_m. To prescribe the density at an arbitrary point ρ in the Γ_c region, we introduce the polar coordinate function $g = g(\rho, \Phi)$ and express the total probing effect $E = E(\rho, \Phi, t)$ as a sum of the incident $E = E(\rho, \Phi, t)$ and secondary $E_S = E_S(\rho, \Phi, t)$ effects: $E = E_i + E_S$. The problem then reduces to the reconstruction of the density distribution across the Γ_c region. Obviously, when a target is probed by electromagnetic radiation, the quantity E_i is the part of E directly related to the initial wave front which is the first to arrive at any point in Γ_c, while E_S is composed of all effects scattered, often repeatedly, by all the points in the Γ_c region.

The secondary probing effect must be given as

$$E_S(\vartheta, t) = \Omega\{g(\rho, \Phi); E(\rho, \Phi, t); w\}, \tag{1.15}$$

where $E_S(\vartheta, t)$ is the amplitude value of $E_S(\rho, \Phi, t)$ in Γ_m, $\Omega\{\ldots\}$ is an integral operator determined in the Γ_c region, and w is the distance between the point P and the receiver B. It is easy to see that a tomographic problem is a classical inverse source problem, since the function $g(\rho, \Phi)$ is to be reconstructed from the known values of $E_S(\vartheta, t)$ and the source in Γ_m. Note that here we are faced with the problem of dimensionality, because $E_S(\vartheta, t)$ measurements are 2D, while the resulting effect $E(\rho, \Phi, t)$ is 3D. Because of this discrepancy, inversion algorithms become numerically unstable and sensitive to any error in the initial data.

The solution to the inverse source problem is always approximate. An approximation most important to tomography involves geometrical optics allowing the representation of probing effects as rays. This provides an optimal formulation of the inverse problem related directly to conventional computerised tomography which reconstructs images from linear trajectories. This can be illustrated with Fig. 1.5. The signal recorded at point B can be represented as a function of the variables ϑ and Φ to show that this signal varies with the position of the point B in Γ_m and with the radiation incidence:

$$S(\vartheta, \Phi) = \int_{l(A)}^{l(B)} g(\rho, \Phi) dl, \tag{1.16}$$

where l is a coordinate going along the ray, whose initial and final points are denoted as $l(A)$ and $l(B)$, respectively.

There are no dimensionality problems with this expression, because the measured quantity $S(\vartheta, \Phi)$ and the reconstructed quantity $g(\rho, \Phi)$ are 2D. So if $S(\vartheta, \Phi)$ is prescribed for the number of ϑ and Φ pairs sufficient for the description of $g(\rho, \Phi)$ with the desired accuracy, the true density distribution may be reconstructed such that the computational algorithm is stable. Equation (1.16) is a governing equation in conventional tomography. At present, there are various reconstruction techniques allowing the solution of this integral equation [88].

No doubt, it would be desirable to integrate the true image in the Γ_c region (in the image space). For practical considerations, however, the data may be integrated in a different space, whose properties depend on how the experimental data are related to the density function $g(\rho, \Phi)$. The quantity to be measured is often a Fourier image of the density distribution, so the data recording is said to be performed in the Fourier space. An example of this type of recording is that in a radio telescope with a synthetic aperture [118]. Although the data integration in the image space and the Fourier space is identical theoretically, the practical algorithms for image reconstruction differ

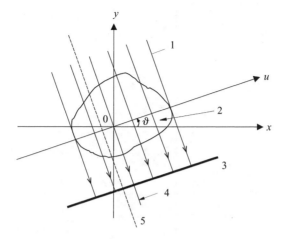

Figure 1.6   A scheme of X-ray tomographic experiment using a collimated beam: 1 – X-rays; 2 – projection angle; 3 – registration line; 4 – projection axis and 5 – integration line

essentially. Many of the available algorithms for the reconstruction of an unknown 2D function g are based on the projection slice theorem [57,95]. It can be formulated with reference to Fig. 1.6 by introducing, in the image space, two rectangular coordinate systems xOy and uOv, rotated by the angle ϑ relative to each other. The projection of the g function at the angle ϑ is described as

$$P_\vartheta(u) = \int_{-\infty}^{\infty} g(u\cos\vartheta - v\sin\vartheta, u\sin\vartheta - v\cos\vartheta)\, dv, \tag{1.17}$$

where $P_\vartheta(u)$ calculated at constant $u = u_o$ is a 1D integral along the respective straight line parallel to the v-axis, so that the $P_\vartheta(u)$ function describes a set of integrals for all ϑ values. The projection theorem states [57] that a 1D Fourier image of a projection made at an angle $P_\vartheta(u)$ represents a 'slice' of a 2D Fourier transform of the $g(x,y)$ function at the ϑ angle to the X-axis:

$$P_\vartheta(U) = G(U\cos\vartheta, U\sin\vartheta) \tag{1.18}$$

with

$$P_\vartheta(U) = \int_{-\infty}^{\infty} P_\vartheta(u)e^{-juU}\, du,$$

$$G(X,Y) = \int_{-\infty}^{\infty}\int g(x,y)e^{-j(xX+yY)}\, dx\, dy.$$

18 Radar imaging and holography

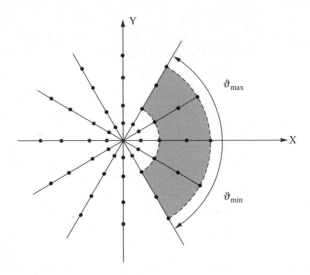

Figure 1.7 *The geometrical arrangement of the G(x, y) pixels in the Fourier region of a polar grid. The parameters ϑ_{max} and ϑ_{min} are the variation range of the projection angles. The shaded region is the SAR recording area*

In classical X-ray tomography, a body is probed by a collimated radiation beam (Fig. 1.6), while the $P_\vartheta(u)$ function is measured by a set of sensors located along a straight line normal to the radiation direction. The set of $P_{\vartheta i}(u)$ projections for various ϑ angles is formed by rotating the object or the power transmitters and receivers. Then one usually uses a convolution back-projection (CBP) algorithm to be discussed later. An alternative is to use a Fourier transform. The latter approach is convenient when data recording is made in the Fourier space and the pixel values of the $P_{\vartheta i}(U)$ Fourier images are known. According to the projection theorem, these pixels also represent the pixels of the $G(X, Y)$ function along a line at the ϑ angle to the X-axis. Therefore, the $P_{\vartheta i}(U)$ values obtained for a set of ϑ angles prescribe the $G(X, Y)$ pixels on a polar grid (Fig. 1.7). By using an interpolation algorithm, one can go over to the $G(X, Y)$ pixels on a rectangular grid and use an inverse Fourier transform to reconstruct the density $g(x, y)$. There have been attempts to compute $g(x, y)$ directly from the $G(X, Y)$ pixels on a polar grid to avoid using an interpolation algorithm [100].

Let us now discuss common approaches used in computerised tomography and radar imaging. In the latter, the target position is determined from the time delay of the radar echo and the antenna orientation. The range resolution is usually much higher than the angular resolution. Suppose a wide beam transmitter and a receiver of electromagnetic radiation are located at the points C and C′, respectively (Fig. 1.5). The geometrical positions of scatterers, whose echo-signals arrive at the point C′ simultaneously, form an ellipse with the focus at C and C′. More exactly, this is a band between two 'concentric' ellipses, its width characterising the resolution limit of the system. Part of this band is denoted as $\gamma-\gamma$. In the case of imaging one point (when the points CПC′ overlap), the ellipses degenerate into circles. The total scattering intensity is proportional to the band-averaged density of scatterers. Since the distance between the ellipses is equal to the resolution, it is sufficient to integrate the density along an average ellipse. The signal recorded at the point C can be written as

$$S(C, C'; \gamma) = \int_\gamma g(\rho, \Phi) \, dl, \tag{1.19}$$

where γ denotes an average ellipse and dl is an element of the ellipse length.

Like in the case described by Eq. (1.16), there is no problem of dimensionality. The scatterers located in the vicinity of a certain point Q can be identified by changing the positions of the transmitter and the receiver. Figure 1.5 shows one of these positions, denoted as D and D', and the respective band δ–δ.

The true density distribution can be reconstructed from a sufficiently large number of measurements made at different points. It is clear that the cases described by Eqs (1.16) and (1.19) differ only in the integration direction.

In X-ray tomography, the function $\rho(x, y)$ describes an unknown distribution of the X-ray attenuation coefficient across a transversal slice (Fig. 1.6) to be measured. The $P_\theta(u)$ projection values are obtained in a multi-beam system represented as an array of X-ray transmitters and receivers located at the θ angle to the x-axis (Fig. 1.6). The intensity of the received radiation decreases exponentially as the beams pass along the line of $\rho(x, y)$ integration. Therefore, the projection $P_\theta(u)$ of this function is

$$P_\theta(u) = -\log \frac{I_\theta(u)}{I_o}, \tag{1.20}$$

where I_o is the X-ray source intensity and $I_\theta(u)$ is the intensity registered by the receivers. The set of $P_{\theta_i}(u)$ projections can be obtained by rotating the object or the array of transmitters and receivers by discrete angles $\theta = \theta_i$. The distribution of the attenuation coefficient of $g(x, y)$ is usually reconstructed from the measured $P_{\theta_i}(u)$ projections, using the CBP method [127]. It enables one to estimate the spatial distribution of the inner physical parameters of the target.

It will be shown below that a radar can register the $P_\theta(v)$ signal. For a given θ angle, its intensity is proportional to the scattering density $\rho_\theta(u, v)$ integrated along the u- and v-coordinates, that is, it is a tomographic projection along the v-axis. So the $P_\theta(v)$ function is a 1D function of the variable v with the parameter θ defining the projection orientation. One can see that there is an essential difference between X-ray tomography and synthetic aperture radar (SAR) imaging. In the latter, a linear integral used to obtain a projection is taken in the direction normal to the microwave propagation, whereas in X-ray tomography it is taken along the X-ray propagation direction.

It will be shown in Chapter 6 that the Doppler and holographic methods of SAR signal processing can provide such projections. Another important specificity is that these projections include a phase factor to describe the time of the double path of the signal between a target and a radar antenna. Thus, a projection produced by SAR is a coherent tomogram that carries much more information about a target. This is especially evident when one uses holographic projections (see Chapter 6). On the other hand, a tomographic processing of projections is capable of reconstructing the arrangement of scatterers on a target, or, in fact, its shape.

1.4 The principles of microwave imaging

The past decade has witnessed an ever increasing interest in radars with a very high resolving power. For example, the ERS-1 and ERS-2 radars (side-looking synthetic aperture radars (SARs) of the European Space Agency [62]) provide microwave imagery of the earth surface with the resolution of 25 m × 25 m in the azimuth-range coordinates. An earth area of 100 km × 100 km (100 km is the radar swath width) is represented by 1.6×10^7 pixels. Modern ground radars have large antenna arrays with an aperture of about 10^4–$10^5 \lambda_1$, where λ_1 is the radar wavelength. They provide an angular resolution of 10^{-4}–10^{-5} rad [129], so the radar vision field can be subdivided into 10^4–10^5 beams.

A radar with a linear and angular resolution much higher than that of a TV equipment ($7.10^5 - 10^6$ pixels) is capable of producing microwave images of extended targets (land areas and water surfaces) and complex objects (aircraft, space craft). So it is reasonable to give a definition of a microwave image. At present, there is no generally accepted definition, so we suggest the following formulation. A microwave image is an optical image, whose structure reproduces on a definite scale the spatial arrangement of scatterers ('radiant' points) on a target illuminated by microwave beams. In addition to the arrangement, scatterers are characterised by a certain radiance. It should be emphasised that the microwave beams can produce 3D images, whereas the visible range of conventional optical systems gives only 2D images. Available methods of microwave imaging can be grouped into three classes:

- direct methods using real apertures;
- methods employing synthetic apertures;
- methods combining real and synthetic apertures.

Imaging by direct methods can, in turn, be performed by real antennas or antenna arrays. Real antennas were used in the early years of radar history. An earth area was viewed by means of circular scanning or sector rocking of the antenna beam in the azimuthal plane. Such systems were termed panoramic or sector circular radars. Modern panoramic radars use 50–100λ_1 apertures and their resolution is low. Since the application of airborne panoramic antenna arrays is a hard task, the only way to increase the resolution is to use the millimetre wavelength range. One is faced with a similar problem when dealing with a side-looking real antenna mounted along the aircraft fuselage. Such antennas may be as long as 10–15 m; at the wavelength $\lambda_1 = 3$ cm, their angular resolution is less than 10 min of arc and the linear resolution of the earth surface is a few dozens of metres, which is too low for some applications. For this type of antenna, the problem of increasing the aperture size was solved in a radical way – by replacing a real aperture with a synthesised aperture.

Consider the potentialities of antenna arrays for aerial survey of the earth surface and for ground imaging of targets flying at low altitudes. Suppose we are to design an antenna array for aircraft imaging. The target has a characteristic size D and is illuminated by a continuous radar pulse. Then, according to the sampling theorem [103], the echo signal function in the aperture receiver can be described by a series of records recorded at the intervals

$$\delta L = \frac{R\lambda_1}{D}, \tag{1.21}$$

where R is the distance to the target. The aperture size necessary for getting a desired resolution on the target Δ, can be defined in terms of Abbe's formula [131]:

$$\Delta = \frac{\lambda_1}{2\sin\alpha/2} \cong \frac{\lambda_1 R}{L}, \tag{1.22}$$

where α is the aperture angle and L is its length.

The total number of receivers on an aperture of length L is

$$N = \frac{L}{\delta L} = \frac{DL}{\lambda_1 R}. \tag{1.23}$$

With Eq. (1.22), we get

$$N = \frac{D}{\Delta}. \tag{1.24}$$

Let us illustrate this with a particular problem. Suppose we have $\lambda_1 = 10$ cm, $R = 600$ km, $D = 20$ m, and $\Delta = 1$ m. Then we get $L = 60$ km, $\delta L = 3$ km and $N = 20$. A planar aperture of $L \times L$ in size must contain $n = N^2 = 400$ individual receivers.

This example shows that the applicability of direct imaging using large antenna arrays is quite limited. Nevertheless, one of these techniques employing a radio camera designed by B. D. Steinberg is of great interest [129]. The radio camera is based on a pulse radar with a real large antenna array and an adaptive beamforming (AB) algorithm. The principal task is to obtain a high resolution with a large aperture avoiding severe restrictions on the arrangement of the antenna elements. The operation of a self-phasing algorithm requires the use of an additional external phase-synchronising radiation source with known parameters, which could generate a reference field and would be located near the target. The radio camera provides an angular resolution of 10^{-4}–10^{-5} rad [129], and the image quality is close to that of optical systems. There is one limitation – the radio camera has a narrow vision field. But still, it may find a wide application in radar imaging of the earth surface, in surveying aircraft traffic, etc.

To summarise, direct real aperture imaging of remote targets at distances of hundreds and thousands of kilometres is practically impossible.

We turn now to methods employing a synthesised aperture. The idea of aperture synthesis born during the designing of a side-looking aperture radar [32,74,86] was to replace a real antenna array with an equivalent synthetic antenna (Fig. 1.8). An antenna with a small aperture is to receive consecutive echo signals and make their coherent summation at various moments of time. For a coherent summation to be made, the radar must also be coherent, namely, it should possess a high transmitter

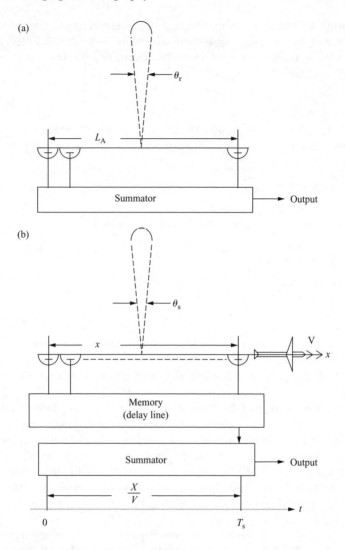

Figure 1.8 Synthesis of a radar aperture pattern: (a) real antenna array and (b) synthesised antenna array

frequency stability and have a reference voltage of the stable frequency to compare echo signals. We shall see below that a reference voltage is similar to a reference wave in holography, with the only difference that the 'wave' is created in the receiver by the voltage of a coherent generator.

Under the conditions described above, the echo signals received by a real antenna are saved in a memory unit as their amplitudes and phases. When an aircraft flies over an earth area $x = L_s$, the signals are summed up at the moment $T_s = x/V$ (the final moment of synthesis), where V is the track velocity of the aircraft. As a result of coherent signal processing, which is similar to the processing by a real antenna (Fig. 1.8(a)), a synthetic aperture pattern θ_s similar to a real aperture pattern θ_r is formed. Thus, the real aperture length L_s is replaced by the synthesised aperture length $x (x = L_s)$. The width of this aperture pattern is

$$\theta_s = \frac{\lambda_1}{2x}. \qquad (1.25)$$

Owing to its large size, a synthetic aperture can provide very narrow patterns, so the track range resolution

$$\delta x = \theta_s R, \qquad (1.26)$$

where R is the slant range to the target, may be very high even at large distances. To illustrate, if the synthetic aperture length is $x = 400$ m and $\lambda_1 = 3$ cm, the resolution may be as high as $\delta x = 6$ m at $R = 160$ km.

Similar principles apply to a stationary ground radar and a moving target. If one needs to obtain a high angular resolution, one can make use of the so-called inverse aperture synthesis. We shall show in Chapter 2 that the resolution on the target is then independent of the distance to it but is determined only by the radar wavelength and the synthesis angle. As a result, one can obtain a very high angular resolution and reconstruct the arrangement of the scatterers into a microwave image.

Thus, current approaches to microwave imaging, based on direct and inverse synthesis of the aperture, provide 2D images which are structurally similar to optical images. Besides, there are methods combining both approaches. They apply a real, say, phased aperture and a synthetic aperture along the aircraft track. These techniques also produce images similar in structure to optical images [2]; they will be discussed in detail in Chapter 3. However, there are certain differences between the two types of 2D images. We summarise the most important ones below.

1. The wavelengths in the microwave range are 10^3–10^6 times longer than in the visible range, and this determines an essential difference in the scattering and reflection by natural and man-made targets. In the visible range, the scattering by man-made targets is basically diffusive, and it can be observed when the surface roughness is of the order of a wavelength. This fact allows a target to be considered as a continuous body. In the microwave range, the picture is quite different because there is no diffusion. The signals are reflected by scattering centres, corner structures and specular surfaces. For this reason, a microwave image of a man-made target is discrete and is made up of 'dark' pixels and those produced by the strong reflectors we mentioned above. A good example is the microwave image of an aircraft that was obtained in Reference 130. Reflection by natural targets produces similar images. However, the reflection spectrum of the earth surface contains an essential diffusion component.
2. For these reasons, the dynamic range of microwave images varies between 50 and 90 dB, while it rarely exceeds 20 dB in optical images, reaching the value of 30 dB in bright sunlight.

3. The quality of an image does not depend on the natural luminosity of a target and depends but slightly on weather conditions.
4. Image quality strongly depends on the geometry of the earth region to be imaged, especially its slant angles, roughness and bulk features in the surface layer. So microwave imaging is used for all-weather mapping, soil classification, detection of boundaries of background surfaces, etc. There is no unified optimal angle (in the vertical plane) for viewing geological structures, and the best values should be adjusted to the local topography. For mountainous and undulated reliefs, for example, a small radiation incidence relative to the normal is preferable, while the imaging of plains requires the use of large incidence angles, which increase the sensitivity to surface roughness. For this reason, images produced by airborne SAR may be inadequate radiometrically (speckle noise) resulting from a large variation in the incidence across a swath because of a wide aperture pattern. Space SARs possess an approximately constant radiation incidence across a swath, so there is no speckle on the image.
5. The density of blackened regions on a negative depends significantly on the dielectric behaviour of the surface being imaged, in particular, on the presence of moisture, both frozen and liquid, in the soil.
6. The microwave range gives the opportunity to probe subsurface areas. For example, the microwave images of the Sakhara desert obtained by a SIR-A SAR showed the presence of dried river beds buried under the sands, which were invisible on the desert surface. This opens up new opportunities to archaeological surveys. It has been demonstrated experimentally that the probing radiation depth in dry sand may be as large as 5 m. Besides, a sand stratum possessing a low attenuation is found to enhance images of subsurface roughness due to refraction at the air–soil interface. This effect is particularly strong for horizontal polarisation at large incidence angles.
7. The specific propagation pattern of the long wavelengths in the microwave range provide quality imagery of lands covered with vegetation.
8. The interaction of subwater phenomena such as internal waves, subsurface currents, etc., with the ocean surface allows imaging the bottom topography and various subwater effects.
9. The use of moving target selection allows one to make precise measurements of the target's radial velocity relative to the SAR.
10. An important factor in imagery is the proper choice of radiation polarisation.
11. Quite specific is imaging of urban areas and other anthropogenic targets. This is due to a large number of objects with a high dielectric permittivity (e.g. metallic objects), surface elements possessing specular reflection, resonance reflectors and objects with horizontal and vertical planes that form corner reflectors. The result of the latter is the following effect: streets parallel to the SAR carrier track produce white lines on the image (the positive), while streets normal to the track produce dark lines. Moreover, the presence or absence on the image of some linear elements of the radar scene and an average density of blackening of the whole image depend on the azimuthal angle, that is, the angle made by the SAR

beam in the plane tangential to the earth surface. This is a serious obstacle to the analysis of images of urban areas.

12. An image contains speckle noise associated with the coherence of the imaging process.

To conclude, a microwave radar image may be 3D if it is recorded by holographic or tomographic techniques (Chapters 5 and 6, respectively).

Chapter 2
Methods of radar imaging

2.1 Target models

All radar targets can be classified into point and complex targets [138]. A point target is a convenient model object commonly used in radar science and practice to solve certain types of problems. It is defined as a target located at distance R from a radar at the viewing point '0', which scatters the incident radar radiation isotropically. For such a target, the equiphase surface is a sphere with the centre at '0'. Suppose a radar generates a wave described as

$$f(t) = a(t) \exp j[\omega_0 t + \Phi(t)],$$

where $f_0 = \omega_0/2\pi$ is the carrier frequency, while a and Φ are the amplitude and phase modulation functions overlapping the carrier frequency.

A point target located at distance R creates an echo signal

$$g(t) = \sigma f\left(t - \frac{2R}{c}\right) = \sigma a\left(t - \frac{2R}{c}\right) \exp j\left[\omega_0\left(t - \frac{2R}{c}\right) + \Phi\left(t - \frac{2R}{c}\right)\right], \tag{2.1}$$

where σ is a complex factor including the target reflectance and signal attenuation along the track.

The Doppler frequency shift is implicitly present in the variable R. If we assume that the radial velocity v_1 is constant, we shall have

$$R = R_1 + v_1 t_1, \tag{2.2}$$

where R_1 is the distance to the target at the initial moment of time $t = 0$.

Equations (2.1) and (2.2) describe a simple model target to be further used for the analysis of the aperture synthesis and imaging principles.

In practice, most radar targets refer to the class of complex targets. In spite of a great variety of particular targets, we can offer a common criterion for their

28 Radar imaging and holography

classification. This criterion is based on the relationship between the maximum target size and the radar resolving power in the coordinate space of the parameters R, α, β and $\dot{R}$, which are the range, the azimuth, the elevation angle and the radial velocity of the target, respectively. An additional important parameter is the number of scattering centres (scatterers). In accordance with this criterion, all complex targets can be subdivided into extended compact targets and extended proper targets. A target is referred to as extended compact if it has a small number of scatterers, its linear and angular dimensions are much smaller than the radar resolution element, and the difference between the radial velocities of the extremal scatterers is appreciably smaller than the velocity resolution element. What is important is that this definition also holds for targets located at large distances. On the other hand, a target which has a size much larger than the radar resolution element and a large number of scatterers should be referred to as extended proper. Earth and water surfaces are examples of such targets.

We shall first discuss extended compact targets (airplanes, spacecrafts, etc.). In the high-frequency region, these targets should be represented as a set of scatterers, or radiant points. The mathematical model of an extended compact target, based on the concept of scatterers, has the form [138]:

$$U = \sum_{m=1}^{M} \sqrt{\sigma_m} \exp(j\Phi_m), \qquad (2.3)$$

where M is the number of individual scatterers, σ_m is the radar cross-section (RCS) for the mth scatterer and Φ_m is the phase of the pulse reflected by the mth scatterer relative to that of the pulse reflected by the first scatterer. The value of σ_m is to be found for a particular polarisation.

Equation (2.3) is usually used for monostatic incidence in the optical region (high frequency approximation). It can also be used to find the relation between monostatic and bistatic scattering at the same target aspect α. For this, the phases of the scatterers, Φ_m, should be expressed as a sum of two terms [69]:

$$\Phi_m = 2kZ_m(\alpha) \cos \beta/2 + \xi_m, \qquad (2.4)$$

where $Z_m(\alpha)$ is the projection of the distance between the mth and the first scatterers onto the bisectrix of the bistatic angle, $k = 2\pi/\lambda_1$ is the wave number of the incident wave and β is the bistatic angle and ξ_m is the residual phase contribution of the mth scatterer, including the contribution of the creeping wave.

For scatterers retaining their position with changing bistatic angle, the mathematical model is

$$U = \sum_{m=1}^{M} \sqrt{\sigma_m} \exp(j2kZ_m(\alpha) \cos \beta/2\xi_m). \qquad (2.5)$$

Equation (2.5) allows us to introduce the concept of equivalence of mono- and bistatic scattering and to define conditions for this equivalence. The theorem of R. E. Kell states that (1) if the total field can be written as a sum of the fields of all scatterers and (2) if the quantity $\sqrt{\sigma_m}$, the Z_m-coordinate and the residual phase ξ_m are all independent of the bistatic angle β in a particular range of β values at any given aspect α, then the total bistatic field for the angles α and β is equal to the monostatic scattering field measured along the bisectrix of the β angle at a frequency reduced by a factor of $\cos(\beta/2)$. This theorem will be used in Chapter 5 to justify the method of inverse aperture synthesis for recording and reconstruction of Fourier holograms.

The amplitude and polarisation characteristics of individual scatterers are of special interest for the understanding of diffraction phenomena in extended compact targets. A comparison of respective experimental and theoretical values should be based on precise scattering models substantiated by the physical theory of diffraction, namely, by the edge wave method (EWM) [137] or by the geometrical theory of diffraction (GTD) [70]. To illustrate, let us consider the field scattered by a perfectly conducting cylinder of finite length l and radius a oriented towards the transmitting antenna. According to the EWM [12], the horizontal and vertical field components in the far range are

$$E_\varphi = \frac{ia}{2} E_{\text{ox}} \frac{e^{ikR}}{R} \overline{\sum}(\vartheta), \qquad E_\vartheta = \frac{ia}{2} H_{\text{ox}} \frac{e^{ikR}}{R} \sum(\vartheta), \qquad (2.6)$$

where k is the wave number and ϑ is the angle between the viewing direction and the cylinder symmetry axis, $\pi/2 \leq \vartheta \leq \pi$:

$$\overline{\sum}(\vartheta) = \overline{\sum}(1) + \overline{\sum}(2) + \overline{\sum}(3), \qquad (2.7)$$

$$\sum(\vartheta) = \sum(1) + \sum(2) + \sum(3), \qquad (2.8)$$

$$\overline{\sum}(1) = f(1)[J_1(\zeta) + iJ_2(\zeta)]e^{ikl\cos\vartheta}, \qquad (2.9)$$

$$\overline{\sum}(2) = f(2)[-J_1(\zeta) + iJ_2(\zeta)]e^{ikl\cos\vartheta}, \qquad (2.10)$$

$$\overline{\sum}(3) = f(3)[-J_1(\zeta) + iJ_2(\zeta)]e^{-ikl\cos\vartheta}, \qquad (2.11)$$

$$\zeta = 2ka\sin\zeta,$$

$J_1(\zeta)$ and $J_2(\zeta)$ are the first- and second-order Bessel functions, respectively. Indices 1, 2 and 3 correspond to three scatterers on the cylinder (Fig. 2.1).

Similar expressions can be obtained for the functions $\sum(1)$, $\sum(2)$ and $\sum(3)$ by replacing $f(1)$, $f(2)$ and $f(3)$ by $g(1)$, $g(2)$ and $g(3)$, respectively. The latter are

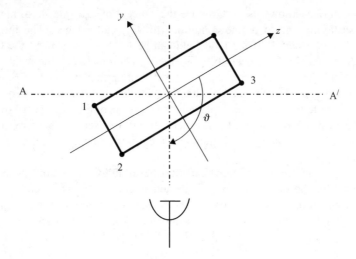

Figure 2.1 Viewing geometry for a rotating cylinder: 1, 2, 3 – scattering centres (scatterers)

defined as

$$\left.\begin{matrix}f(1)\\g(1)\end{matrix}\right\} = \frac{\sin(\pi/n)}{n}\left[\left(\cos\frac{\pi}{n}-1\right)^{-1} \pm \left(\cos\frac{\pi}{n}-\cos\frac{(\pi-2\vartheta)}{n}\right)^{-1}\right], \quad (2.12)$$

$$\left.\begin{matrix}f(2)\\g(2)\end{matrix}\right\} = \frac{\sin(\pi/n)}{n}\left[\left(\cos\frac{\pi}{n}-1\right)^{-1} \mp \left(\cos\frac{\pi}{n}-\cos\frac{2\vartheta}{n}\right)^{-1}\right], \quad (2.13)$$

$$\left.\begin{matrix}f(3)\\g(3)\end{matrix}\right\} = \frac{\sin(\pi/n)}{n}\left[\left(\cos\frac{\pi}{n}-1\right)^{-1} \mp \left(\cos\frac{\pi}{n}-\cos\frac{(\pi+2\vartheta)}{n}\right)^{-1}\right], \quad (2.14)$$

$n = 3/2$.

The functions (2.7)–(2.14) can be used to calculate the scattering characteristics (the RCS diagram, the amplitude and phase scattering diagrams, etc.) for an experimental study of diffraction in an anechoic chamber (AEC). The last two diagrams, for example, can be found as the modulus and argument of the functions (2.7) and (2.8). However, the representation of the field as a sum of the fields re-transmitted by scatterers provides information on individual scatterers. Such characteristics are referred to as local responses [12]. The RCS diagrams for scatterers on a cylinder and the E- and H-polarisations of the incident field can be written as

$$\sigma_n^E(\vartheta) = \pi a^2 \left|\overline{\sum_n}(\vartheta)\right|^2, \quad \sigma_n^H(\vartheta) = \pi a^2 \left|\sum_n(\vartheta)\right|^2, \qquad (2.15)$$

$$n = 1, 2, 3.$$

The phase responses of scatterers can be derived in the form of arguments of the complex valued functions (2.9)–(2.11). A scattering model for a cylinder with bistatic incidence was designed in Reference 12 in the EWM approximation. Besides, it is shown in References 105 and 109 that the amplitude responses and the positions of scatterers on a target can be studied experimentally using images reconstructed from microwave holograms.

We now turn to models of extended proper targets. Such targets include

- land surface;
- sea surface;
- large anthropogenic objects like urban areas and settlements;
- special standard objects for radar calibration.

An analysis of models of all of these targets would go far beyond the scope of this book. We give a brief survey of scattering models of sea surface in Chapter 4, including a model of a partially coherent extended proper target, which is used in the analysis of microwave radar imagery.

It should be noted that extended compact targets may also be partially coherent (Chapter 7). In either case, these targets produce parasitic phase fluctuations which perturb radar imaging coherence.

Target models are used for several purposes: to justify the principles of inverse aperture synthesis, to interpret microwave images, to obtain local RCS of scatterers on standard objects, and to calibrate measurements made in AECs.

2.2 Basic principles of aperture synthesis

We have mentioned in Chapter 1 that the use of a synthetic aperture is necessary if one needs to obtain a high angular resolution of targets at large distances. It has been shown by some researchers [73,109] that the aperture synthesis is, in principle, possible for any form of relative motion of a target and a real antenna; what is important is that the target aspect should change together with the relative displacement.

Today there are two basic methods of aperture synthesis – direct and inverse. Direct synthesis can be made by scanning a relatively stationary target by a real antenna (Fig. 2.2(a)). The target is on the earth surface and the antenna is located on an aircraft. Radar systems with direct antenna synthesis are known as side-looking synthetic aperture radars (SARs). The authors of Reference 85 have suggested for them the term quasi-holographic radars (Chapter 3). Methods of aperture synthesis using linear translational motion of a target or its natural rotation relative to a stationary ground antenna are called inverse methods and radars based on such methods are

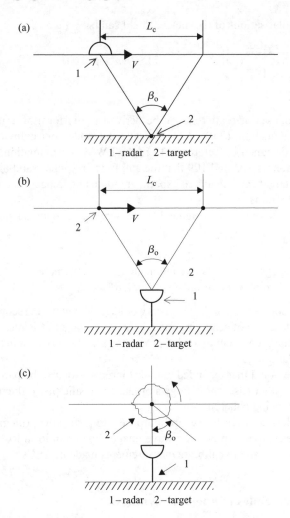

Figure 2.2 Schematic illustrations of aperture synthesis techniques: (a) direct synthesis implemented in SAR, (b) inverse synthesis for a target moving in a straight line and (c) inverse synthesis for a rotating target

known as inverse synthetic aperture radars (ISARs) (Fig. 2.2(b) and (c)). There are also combined approaches to field recording. For example, a front-looking holographic radar (Chapter 3) combines direct synthesis along the track and transversal synthesis with a one-dimensional (1D) real antenna array (Fig. 3.4). A spot-light mode of synthesis is also possible: it uses both the linear movement of an airborne antenna and its constant axial orientation to a ground target (Fig. 3.11). Radars based on this principle are known as spot-light SAR [100].

Finally, ground radars operating in the inverse synthesis mode and viewing a linearly moving target can combine a real-phased antenna array and aperture synthesis.

This method was suggested by B. D. Steinberg [129] to employ adaptive beamforming (AB) together with aperture synthesis (ISAR + AB).

In any method of aperture synthesis, the radar azimuthal resolution is determined by the aperture angle $\beta_o = L_s/R$. The linear resolution along the angle coordinate is $\delta l = \lambda_1/\beta_o$. It should be emphasised [73] that rotation of a synthetic antenna pattern (SAP) does not shift the target phase centre and, therefore, does not synthesise the aperture. For this reason, one cannot increase the angular resolution by rotating a real antenna, in contrast to the target rotation.

2.3 Methods of signal processing in imaging radar

Imaging radar signal processing can be considered from different points of view. Since there is an essential difference between the direct and inverse modes of aperture synthesis, the processing techniques should be described individually for each type of radar.

2.3.1 SAR signal processing and holographic radar for earth surveys

The SAR aperture synthesis by coherent integration is treated in terms of

- the antenna approach [74];
- the range-Doppler approach [85,140];
- the cross-correlation approach [85];
- the holographic approach [85,143];
- the tomographic approach [100].

The use of a variety of analytical techniques in radar imaging leads to various processing designs and physical interpretations of some of its details.

The first four approaches provide a fairly complete analysis of the effects of SAR parameters on its performance characteristics and the results are generally consistent. Each approach, however, enables one to see the image recording and reconstruction in a new light, because each has its own merits and demerits. In this book, we largely follow the holographic approach to the performance analysis of various SAR systems, which involves the theories of optical and holographic systems. According to one of the pioneers of optical and microwave holography E. H. Leith, a holographic treatment of SAR performance has proved most fruitful. The recording of a signal is regarded as that of a reduced microwave hologram of the wave field along the azimuth, that is, along the flight track. Illumination of such a hologram by coherent light reconstructs the optical wave field, which is similar to the recorded microwave field on a certain scale. A schematic diagram illustrating the holographic approach to SAR signal recording and processing is presented in Fig. 2.3. For a point target, for instance, an optical hologram is a Fresnel zone plate. When the plate is illuminated by coherent light, the real and virtual images of the point target are reconstructed (Fig. 3.1).

Thus, a microwave image of a point target can be obtained directly owing to the focusing properties of a Fresnel zone plate. The processing optics in that case is

34 Radar imaging and holography

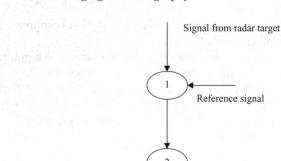

Figure 2.3 The holographic approach to signal recording and processing in SAR: 1 – recording of a 1D Fraunhofer or Fresnel diffraction pattern of target field in the form of a transparency (azimuthal recording of a 1D microwave hologram), 2 – 1D Fourier or Fresnel transformation, 3 – display

necessary only to compensate for various distortions inherent in SAR; anamorphism and the difference in the azimuth and range scale factors. Optical processing of SAR signals was first analysed in terms of holography [86].

The holographic approach will be used in Chapter 3 to describe SAR as a system for combined recording and reconstruction of microwave holograms. A general scheme of this process is shown in Fig. 2.3. The reference signal here is a heterodyne coherent pulse, whose role is actually much more important (see below).

Holographic SAR for surveying the earth surface (Chapter 3) uses the cross-correlation [72] and holographic approaches. The scheme illustrating the holographic principle is similar to that in Fig. 2.3 with the only difference that one deals here with 2D microwave holograms.

The tomographic approach is applied in descriptions of aperture synthesis by spot-light SAR (Chapter 3).

2.3.2 ISAR signal processing

Methods of inverse aperture synthesis have been discussed in a number of publications. The treatments involved are:

- a range-Doppler algorithm [13,21,24];
- a circular convolution algorithm (CCA) [94];

- correlated processing [13];
- extended coherent processing (ECP) [13];
- polar format processing [13];
- holographic processing [109];
- tomographic processing [9,106].

A serious limitation of the range-Doppler algorithm is its applicability only to a synthesis made at relatively small angle steps, which is an obstacle in achieving high resolutions. The restrictions on the time intervals of coherent processing were formulated by D. A. Ausherman *et al.* [13]. Any attempt to overcome these restrictions leads to displacement of individual scatterer images into adjacent resolution elements and, hence, to the image degradation. The range-Doppler algorithm has been used in SAR for microwave imaging of aircraft [8]. Preliminarily, the radial movement of the target is compensated for in all range channels. The development of new processing algorithms based on larger angle steps required the use of spherical coordinates (polar coordinates in the 2D case) instead of the Cartesian coordinates of the range-cross range type. One of these is the CCA permitting the limit angle step of 2π with a precise aperture focusing over the whole target space [94]. Moreover, it is applicable to the processing of both narrow- and wide-band radar signals. The conditions for viewing real targets differ from the conditions, in which the CCA operates. First, discrete records for the angle steps of the target aspect variation, recorded at a constant repetition frequency, are not equidistant. Second, the angle between the radar line of sight (RLOS) and the rotation axis changes during the viewing. The first obstacle can be bypassed by interpolating the radar data. The second one inevitably leads to the necessity to consider a 3D problem. Attempts at using this algorithm, like other 2D algorithms, to process 3D data result in distorted images [8].

When applied to narrow-band signals, the CCA has another disadvantage: the whole ensemble of radar data must be processed simultaneously. So this algorithm should be employed only in measuring test areas and in AECs.

Correlated processing provides well-focused images of targets of any size, and the time intervals of coherent processing may be of arbitrary duration. On the other hand, its computational efficiency is quite low [8].

Both algorithms require special measures to compensate for the phase shift due to the radial displacement of the target.

Extended coherent processing is based on coherent summation of microwave images, each of which is formed by a range-Doppler algorithm at a small angle step. The application of this technique increases the processing rate by approximately an order of magnitude with a good image quality for a fairly long processing time.

Variable movement of a target relative to the RLOS necessitates the use of different algorithms for the synthesis of the final image from partial ones. So algorithms for ECP are subdivided into those for wide angle imaging and those for multiple target rotations. Target aspects suitable for wide angle imaging are chosen when a ground radar views a space craft stabilised along three axes or rotating around its centre of mass. Imaging by multiple rotations has the following specificity: when a space target is stabilized by rotation, the angle step remains the same in every consecutive rotation of the target around its axis. In its latter modification, the ECP algorithm is used for 3D and stroboscopic microwave imaging [13].

Polar format processing is another effective way to overcome the scatterers' movement through the resolution elements. It is based on the representation of radar data in a 3D frequency space.

In our opinion, a very perspective way of inverse aperture synthesis is by holographic processing [109,146]. The possibility of using a holographic approach was first suggested by E. N. Leith [85]. Not only does it provide a new insight into the processes occurring in inverse synthesis but it also helps to find novel designs of recording and reconstruction devices.

The schematic diagram of the holographic approach to ISAR signal recording and reconstruction is similar to that shown in Fig. 2.3. The first step is to record a 1D quadrature or complex microwave Fourier hologram (the diffraction pattern of the target field) (Section 2.3.1). The reference signal is a coherent heterodyne pulse. The second step is the implementation of a 1D Fourier transform. The next step is the image representation.

Tomographic processing can be performed using one of the three ways of image reconstruction:

- reconstruction in the frequency region [9];
- reconstruction in the space region by using a convolution back-projection algorithm [9];
- reconstruction by summation of partial images (Chapter 6).

The tomographic approach to ISAR analysis will be discussed in Section 2.4.2 and in Chapter 6.

2.4 Coherent radar holographic and tomographic processing

2.4.1 The holographic approach

Direct hologram recording commonly used in the optical wavelength range finds a limited application in the microwave range because of the absence of a suitable substitution to a microwave photoplate. So the processing can be made by either of the two methods – direct or inverse aperture synthesis (Section 2.2). These techniques allow the recording of two types of hologram. One is similar to an optical hologram, while the other has no optical counterpart.

Suppose $a \exp(i\Phi)$ is a target wave and $a_o \exp(i\Phi_o)$ is a reference wave. In the first case a square microwave hologram is formed which is described by the following equation:

$$H_1(x,y) = |a\exp(i\Phi) + a_o \exp(i\Phi_o)|^2 = a_o^2 + a^2 + 2aa_o \cos(\Phi - \Phi_o), \tag{2.16}$$

Such holograms can be recorded by a quadratic detector in the high- and medium-frequency ranges (Fig. 2.4(a) and (b) respectively), using a high-frequency reference

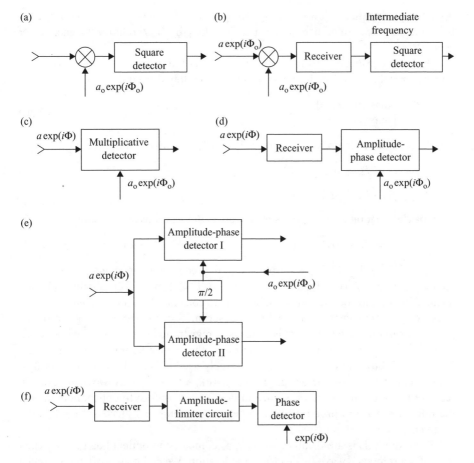

Figure 2.4 Synthesis of a microwave hologram: (a) quadratic hologram recorded at a high frequency, (b) quadratic hologram recorded at an intermediate frequency, (c) multiplicative hologram recorded at a high frequency, (d) multiplicative hologram recorded at an intermediate frequency, (e) quadrature holograms, (f) phase-only hologram

wave. In the second case a multiplicative hologram is formed [109] which is defined as

$$H_1(x,y) = \text{Re}[a\exp(i\Phi) \cdot a_o \exp(-i\Phi_o)] = aa_o \cos(\Phi - \Phi_o). \tag{2.17}$$

The latter can also be formed at high and medium frequencies (Fig. 2.4(c) and (d), respectively).

In either case it is possible to record a quadrature microwave hologram

$$H_2(x,y) = a_o^2 + a^2 + 2aa_o \sin(\Phi - \Phi_o), \tag{2.18}$$

$$H_2(x,y) = aa_o \sin(\Phi - \Phi_o). \tag{2.19}$$

A pair of quadrature microwave holograms (2.16), (2.17) or (2.18), (2.19) is recorded by using identical reference waves phase-shifted by $\pi/2$ relative to each other (Fig. 2.4(e)).

Optical recording of the bipolar functions (2.17) and (2.19) for optical reconstruction requires the use of the reference level H_r to be found from the condition

$$H_r \geq \begin{cases} \max |H_1(x,y)| \\ \max |H_2(x,y)| \end{cases} \qquad (2.20)$$

and the linearity of the microwave recording. Then we arrive at the equations

$$H_1(x,y) = H_r + aa_o \cos(\Phi - \Phi_o), \qquad (2.21)$$

$$H_2(x,y) = H_r + aa_o \sin(\Phi - \Phi_o). \qquad (2.22)$$

Each pair of quadrature holograms makes up a complex microwave hologram:

$$H(x,y) = H_1(x,y) + iH_2(x,y). \qquad (2.23)$$

The quantity j in Eq. (2.23) is introduced at the reconstruction stage, following the recording of only two quadrature holograms, say, (2.21) and (2.22). But this form of a complex hologram equation makes it possible to consider this pair as an entity, which is especially convenient for an analytical description of the reconstruction process. A complex microwave hologram is a means of registration of the total field scattered by a target. It will be shown later that this allows the reconstruction of a single image. The designs shown in Fig. 2.4(a)–(d) are largely used in laboratory and test set-ups, while radar stations use the design in Fig. 2.4(e). A typical microwave holographic receiver based on this design is shown in Fig. 2.5. In contrast to optical holography, the reference wave is produced by a coherent generator and phase-shifter 1 in the receiver.

Therefore, this is a radically new way, as compared to optical holography when it creates a reference wave by electrical modulation. We call it an artificial reference wave. Its incidence angle can be simulated by varying the phase with phase-shifter 1 operating synchronously with the movement of the real radar antenna. The incidence angle α to the carrier track (Fig. 2.6) can be simulated by changing its phase as

$$\Psi = \frac{2\pi x \sin \alpha}{\lambda_1}, \qquad (2.24)$$

where x is the position of the real antenna during the aperture synthesis.

Microwave holograms can be classified in terms of the volume of recorded data on the target wave. If a hologram contains data on the wave amplitude and phase, it is said to be an amplitude–phase hologram. If the amplitude factor $a(x,y)$ is neglected before the summation or multiplication of the target and reference waves, a hologram is said to be a phase-only hologram [109] (Fig. 2.4(f)).

To describe the fields of reconstructed images, one can conveniently use the Fresnel–Kirchhoff diffraction formula [121] employed in optical holography. So it is reasonable to classify holograms in terms of the phase fronts of fields induced by reference sources and diffracted by a target.

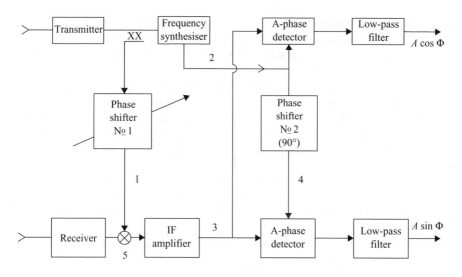

Figure 2.5 A block diagram of a microwave holographic receiver: 1 – reference field, 2 – reference signal $\cos(\omega_0 t + \varphi_0)$, 3 – input signal $A\cos(\omega_0 t + \varphi_0 - \varphi)$, 4 – signal $\sin(\omega_0 t + \varphi_0)$ and 5 – mixer

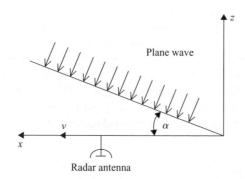

Figure 2.6 Illustration for the calculation of the phase variation of a reference wave

A Fresnel microwave hologram is synthesised by registration of the interference pattern of interaction between plane or spherical reference waves and waves diffracted by a target, which have a spherical phase front in the hologram plane.

A Fraunhofer microwave hologram is formed by recording the interference pattern of plane or spherical reference waves interacting with diffracted waves having a plane phase front in the hologram plane.

A Fourier microwave hologram is formed by recording the interference pattern of interaction between the diffracted waves having a spherical front in the hologram plane and a spherical reference wave with a curvature radius equal to an average curvature radius of the waves coming from the target and propagating in the same direction.

Fresnel and Fraunhofer holograms have found application in SAR theory, while Fourier holograms are used in ISAR theory (Chapters 3 and 5).

Since the process of hologram synthesis implies that the radar is to be coherent, the question arises as to what requirements must be imposed on the coherence. Let us first define the concept of coherence in microwave radar theory. A signal is said to be coherent if it shows no abrupt changes in the basic frequency, or if such changes are small, of the order of 1–3° [14]. If the basic frequency changes are greater than these values, the signal reflected from a target is called partially coherent. This happens when the coherence is perturbed due to:

- an unstable frequency of the radar wave synthesiser or heterodyne;
- the effects of the target itself, say, of a sea surface (Chapter 4);
- a non-uniform motion of the aircraft, for example, yawing, pitching and beaking (Chapter 7);
- the effects of the troposphere and ionosphere, such as sporadic changes in the wave propagation conditions (Chapter 8).

Within this definition, a continuous radiation is always coherent for a period of time when various instabilities in the transmitter performance can be neglected. When a radar operates in a pulse mode, coherence is determined by an unambiguous relation between the initial phase values of the carrier frequency of a train of pulses. The above definition of coherence also applies to radar signals with known phase jumps that can be avoided using coherent sensing. Since the first of the factors responsible for coherence instability is the most serious one, there was a suggestion to introduce in imaging theory the concept of frequency, rather than coherence, stability [87]. A comprehensive analysis of requirements on the frequency stability was made in SAR theory by R.O. Harger [55]. A simplified approach is considered in Reference 87. The latter will be discussed here in more detail in order to explain the physical mechanism of SAR instability. The treatment of this problem has yielded the following expression:

$$\pi \alpha T^2 \leq (\pi/4)(cT/2R), \tag{2.25}$$

where α is the rate of linear frequency variation due to the instability of the radar generator, T is the time for a pulse to reach the target at distance R and to come back. It is clear from Eq. (2.25) that a permissible phase error $\pi \alpha^2 T^2$ is $\pi/4$ for the time $T = 2R/c$.

Therefore, Eq. (2.25) is the criterion for the coherence length in the holographing of reflecting targets; it should provide the frequency stability of the signal propagation for a time consistent with the scene depth (a full analogy with optical holography). Similar stability requirements can be imposed on coherent ISAR, in which coherence is preserved if the signal phase deviation due to the frequency instability is less than $\pi/2$. Then we have the expression

$$2\pi \delta f_c T \leq \pi/2, \tag{2.26}$$

where δf_c is the deviation of the probing signal frequency for the time T. Neglecting the signal delay in the antenna-feeder waveguide, we get

$$\delta f_c \leq c/8R, \tag{2.27}$$

using the concept of short-term instability

$$\varepsilon_f = \frac{\Delta f_c}{f_c}, \tag{2.28}$$

where f_c is the radar carrier frequency. Then we have

$$\varepsilon_f = \frac{c}{8 f_c R}. \tag{2.29}$$

The condition for a long-term frequency instability can be found from a coherent processing in the whole time interval of the synthesis, T_s, which varies with the type of hologram processing.

The frequency stability in modern radars is achieved with highly stable, mainly caesium atomic beam standards of time and frequency. The frequency standards provide a long-term instability over 1 h with a possible adjustment of about 10^{-12}–10^{-14} and a 1 ns random component of the 24 h behaviour of the timescale [25]. To maintain the stability, modern radars use a phase loop control [44]. The long-term instability requirements to coherent ISAR are very high. For example, in the Goldstone Solar System Radar (GSSR) radar for planet surveys [44] this parameter is about 10^{-15} for 1000 s and the pulse-to-pulse instability is less than $1°$. The GSSR Project is designed for the observation of Mercury, Venus and Mars. In a LRIR (Long-Range Imaging Radar), the pulse-to-pulse instability is about 2–$3°$ [20]. This radar is designed for observation of space objects.

To summarise the discussion of factors causing coherence instabilities in radars, the frequency stability maintained by frequency standards and loop frequency control can solve the problem of operation instabilities of a pulse generator or heterodyne. The other three causes of instability can be removed by special signal processing in the radar (Chapters 4 and 7).

2.4.2 Tomographic processing in 2D viewing geometry

It has been shown above that the signal processing in ISAR can be described in terms of Doppler frequencies, correlated processing, CCAs, etc. We believe that the most appropriate approach is tomographic processing which allows focusing of a synthesised aperture over the whole target space and provides an image resolution restricted only by the diffraction limit [7,9,10]. Another advantage of this technique is great possibilities for optimisation of processing algorithms and devices. Consider a target being probed by a stationary coherent radar (Fig. 2.7), which radiates pulses with the carrier frequency f_c and the modulation function $w(t)$

$$s(t) = w(t) \exp(j 2\pi f_c t) \tag{2.30}$$

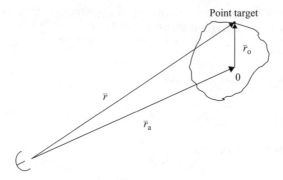

Figure 2.7   The coordinates used in target viewing

and measures the amplitude and phase of the complex envelope of an echo signal. The target is assumed to consist of a small number of independent scatterers, whose position relative to the centre of mass of the target O and the radar is defined by the respective vectors (Fig. 2.7). The target moves along an arbitrary trajectory, rotating around its centre of mass. The conditions for the far zone and a uniform field amplitude of the wave incident on the target surface facing the radar are fulfilled. The algorithm for the processing of the complex envelope of an echo signal, synthesised by the radar receiver, is

$$s_V(t) = \int_V g(\bar{r}_o) w\left(t - \frac{2|\bar{r}|}{c}\right) \exp(-j2k_c|\bar{r}|) d\bar{r}_o, \tag{2.31}$$

where $g(\bar{r}_o)$ is the function of the target reflectivity and $k_c = 2\pi/\lambda_c$ is a wave number corresponding to the wavelength of the radar carrier oscillation. Equation (2.31) allows the estimation of the $\hat{g}(\bar{r}_o)$ reflectivity of every scatterer.

The integration of Eq. (2.31) is made over the target space. With the condition for the far zone, the vector $\bar{r}$ describing the position of an arbitrary scatterer relative to the radar can be substituted by its projection on the line of sight:

$$|\bar{r}| = |\bar{r}_a| + \hat{r}, \tag{2.32}$$

where

$$\hat{r} = \bar{r}_o \bar{u}, \qquad \bar{u} = \bar{r}_a/|\bar{r}_a| \tag{2.33}$$

and $\bar{u}$ is a unit vector coinciding with the line of sight and directed away from the target rotation centre towards the radar.

Generally, both terms of Eq. (2.32) vary during the viewing. However, the contribution to the imaging is made only by the variation in the relative range $\hat{r}$. On the contrary, the range variation of the target's centre of mass $|\bar{r}_a|$ produces distortions in the image. By substituting Eq. (2.32) into Eq. (2.31) and regrouping the terms for the complex envelope distortion, we obtain

$$s_v(t) = \int_V g(\bar{r}_o) \left\{ w\left(t - \frac{2|\bar{r}_a|}{c} - \frac{2\hat{r}}{c}\right) \right\} \exp(-j2k_c|\bar{r}_a|) \exp(-j2k_c\hat{r}) d\bar{r}_o. \tag{2.34}$$

It follows from the analysis of Eq. (2.34) that the correction of the received signal is to maintain a constant delay $\tau = 2|\bar{r}_a|/c$ and to multiply the signal by the phase factor $\exp(j2k_c|\bar{r}_a|)$. After making the correction, the signal can be written (assuming $\tau = 0$) as

$$s_v(t) = \int_V g(\bar{r}_o) w\left(t - \frac{2\hat{r}}{c}\right) \exp(-j2k_c\hat{r}) d\bar{r}_o. \tag{2.35}$$

The exponential phase factor in Eq. (2.35) defines the coherence degree of the whole imaging system (the radar and the processing system) over the whole band-limited frequency spectrum. The coherence instability due to, say, an inaccurate compensation for the target radial movement leads to a poorer resolution. The possibility of imaging by a tomographic algorithm is, in principle, preserved. Let us process a signal in the frequency domain. The Fourier transform of a video signal corresponding to the change in the target aspect relative to the radar is

$$S(f) = F\{s(t)\} = W(f) \int_V g(\bar{r}_o) \exp[-j2(k_c + k)\hat{r}] d\bar{r}_o, \tag{2.36}$$

where $W(f) = F\{w(t)\}$ is the modulation function spectrum, $k = 2\pi/\lambda$ is the wave number to be defined in the frequency spectrum, and $F\{\cdot\}$ is a 1D Fourier transform operator. Next, we perform a standard range processing to obtain the resolution along the line of sight in a filter with the transmission characteristic $K(f)$ [18]:

$$S(f) = H(f) \int_V g(\bar{r}_o) \exp[-j2(k_c k)\hat{r}] d\bar{r}_o \tag{2.37}$$

with $H(f) = W(f)K(f)$.

The range processing can also be made in the time domain of the receiver using a filter with the impulse response $h(t) = F^{-1}\{K(f)\}$, where $F^{-1}\{\cdot\}$ is the inverse Fourier transform (IFT) operator.

Note that the compensation for the target radial displacement can also be made by the processor (after the transformation of Eq. (2.37)) by multiplying the video signal spectrum by the phase factor $\exp[j2(k_c + k)|\bar{r}_a|]$. A particular method of processing requires a proper design of the receiver and processor.

With Eq. (2.33), expression (2.37) can be presented as a 3D Fourier transform of the target reflectivity:

$$S(f) = H(f) \int_V g(\bar{r}_o) \exp[-j2(k_c + k)\bar{u}\bar{r}_o] d\bar{r}_o, \tag{2.38}$$

where $(k_c + k)$ is the 3D frequency vector modulus.

To calculate the target reflectivity, it is necessary to make an inverse transformation of the Fourier function over the respective volume:

$$\hat{g}(\bar{r}_o) = F^{-1}\{S(f)\} = g(\bar{r}_o) * h(\bar{r}_o), \qquad (2.39)$$

where $*$ denotes convolution, $h(\bar{r}_o)$ is the processing system response from a single point target in the space frequency domain, $h(\bar{r}_o) = F^{-1}\{H(f)\}$, and $H(f)$ is a 3D aperture function.

It is clear from Eq. (2.39) that the value of $\hat{g}(r_o)$ is a distorted representation of the target reflectivity $g(r_o)$. The distortion is largely due to the limited frequency spectrum and the small angle step of the aspect variation.

Equation (2.39) can be transformed in the 3D frequency domain. More often, however, one needs 2D images, which can be obtained using an appropriate 2D data acquisition design (Fig. 2.8). Equation (2.38) then has the form:

$$S(f) = H(f) \iint_{-\infty}^{\infty} g(u,v) \exp[-j2(k_c+k)v]\, du\, dv. \qquad (2.40)$$

Keeping in mind that the function

$$P_\theta(v) = \int_{-\infty}^{\infty} g(u,v)\, du, \qquad (2.41)$$

represents the projection of the target reflectivity on the v-axis, the target aspect defined by the angle θ (Fig. 2.8) can be written as

$$S_\theta(f) = H(f) \int_{-\infty}^{\infty} P_\theta(v) \exp[-j2(k_c+k)v]\, dv. \qquad (2.42)$$

Using the denotation $f_p = 2(f_c+f)/c$, we get

$$S_\theta(f) = H(f)P_\theta(f_p) = H(f) \int_{-\infty}^{\infty} P_\theta(v) \exp(-j2\pi f_p v)\, dv, \qquad (2.43)$$

where $P_\theta(f_p)$ is the Fourier transform of the projection $P_\theta(v)$ with the space frequency f_p.

The substitution of Eq. (2.41) into Eq. (2.43) yields

$$P_\theta(f_p) = \iint_{-\infty}^{\infty} g(u,v) \exp[-j2\pi(0u+f_p v)]\, du\, dv \qquad (2.44)$$

or

$$P_\theta(f_p) = P_\theta(0,f_p) = P_\theta(f_p \sin\theta, f_p \cos\theta), \qquad (2.45)$$

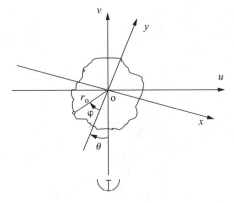

Figure 2.8   2D data acquisition design in the tomographic approach

where $P(\cdot)$ is the Fourier transform of the target reflectivity in the (x,y) coordinates. Then using Eq. (2.45), we have

$$S_\theta(f) = H(f)P_\theta(f_p \sin\theta, f_p \cos\theta). \tag{2.46}$$

Equation (2.45) represents the formulation of the projection theorem underlying the tomographic imaging algorithms [34,57].

Bearing in mind that $v = y\cos\theta - x\sin\theta$, we go from Eq. (2.43) to the 2D Fourier transform in the (x,y) coordinates related to the target:

$$S(f_x, f_y) = H(f) \int\!\!\!\int_{-\infty}^{\infty} g(x,y)\exp[-j2\pi(f_x x + f_y y)]\,dx\,dy, \tag{2.47}$$

where f_x and f_y are the respective space frequencies, $f_x = -(f_{po} + f_p)\sin\theta$, $f_y = (f_{po} + f_p)\cos\theta$, $f_{po} = 2f_c/c$ is the space frequency corresponding to the carrier frequency spectrum, f_p is the space frequency defined over the whole frequency band of the probing signal, $2f_l/c < f_p < 2f_u/c$, f_l and f_u are the lower and top frequency spectra. The solution to Eq. (2.47) yields the target reflectivity:

$$\hat{g}(x,y) = \int\!\!\!\int_{-\infty}^{\infty} S(f_x, f_y)\exp[j2\pi(f_x x + f_y y)]\,dx\,dy = g(x,y) * h(x,y). \tag{2.48}$$

This approach to imaging can be implemented in the frequency and space domains (see Chapter 6). Note that the radar data on a signal are recorded in polar coordinates [8], while the imaging devices are represented as a dot matrix. This inconvenience necessitates the use of a cumbersome procedure of data interpolation and then finding a compromise between the degree of interpolation complexity (the greater the complexity, the better the image quality) and the computation resources. It will be shown

46 Radar imaging and holography

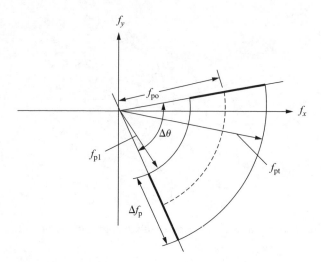

Figure 2.9 *The space frequency spectrum recorded by a coherent (microwave holographic) system. The projection slices are shifted by the value f_{po} from the coordinate origin*

in Chapter 6 that there is a procedure of processing in the space domain, which successfully overcomes this difficulty.

The space spectrum of each echo signal is represented in the frequency $f_x f_y$ plane (Fig. 2.9) as a straight line coinciding with a radial beam. The beam angular coordinate θ is equal to the angle ϑ which defines the target position at the moment of probing signal reflection. The space spectra of echo signals are centred relative to an arc of radius f_{po}. In the frequency plane, their multiplicity forms microwave holograms with the angle $\Delta\theta$ equal to the angle step of the synthesis, $\Delta\vartheta$. The inner and outer radii of a hologram are defined by the space frequencies f_{pl} and f_{pt} in the lower and top frequency spectra of the probing signal.

It follows from Eq. (2.47) that the ensemble of radar data recorded under the above conditions is a 2D Fourier microwave hologram. The image reconstruction from such a hologram reduces to IFT. Although the inversion of a hologram described by Eq. (2.47) is a simple mathematical procedure, the methods of its digital implementation are not as obvious.

With the above assumptions of the far zone and the high-frequency spectrum, we can suggest that at every moment of time $t = 2v/c$ the contribution to the echo signal will be made only by the local scatterers with the range coordinate ϑ. Then the integral

$$P_\vartheta(\vartheta) = \int_V g(u,\vartheta) du \qquad (2.49)$$

taken along the transverse range represents the projection of the target reflectivity on the RLOS.

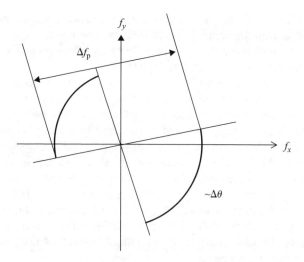

Figure 2.10   *The space frequency spectrum recorded by an incoherent (tomographic) system*

With Eq. (2.49), expression (2.40) will have the form:

$$S_\theta(f_{po}+f_p) = H(f_p)\int_V P_\vartheta(\vartheta)\exp[-j2\pi(f_{po}+f_p)\vartheta]d\vartheta = H(f_p)P_\theta(f_{po}+f_p),$$
(2.50)

where

$$P_\theta(f_{po}+f_p) = F\{P_\vartheta(\vartheta)\exp(-j2\pi f_{po}\vartheta)\}.$$
(2.51)

If the right-hand side of Eq. (2.50) is expressed in the Cartesian coordinates, we shall have

$$S_\theta(f_p) = H(f_p)P_\theta[-(f_{po}+f_p)\sin\theta, (f_{po}+f_p)\cos\theta].$$
(2.52)

Hence, a 1D spectrum of the product of the reflectivity function projection at the angle ϑ and the phase factor $\Phi = \exp(-j2\pi f_{po}\vartheta)$ is the cross section of a microwave hologram function along a straight line passing through the frequency plane origin at the angle θ ($\theta = \vartheta$).

If the data acquisition system is incoherent and records only the complex envelope shape of the echo signal, the phase factor Φ vanishes from Eq. (2.50) to Eq. (2.52). Equation (2.52) then reduces to the projection slice theorem, one of the fundamental theorems in computerised tomography [57].

Let us discuss the physical differences between coherent (holographic) and incoherent (tomographic) systems of microwave radar imaging by comparing Figs 2.9 and 2.10. The angle step of the target aspect variation and the frequency band width of the probing signal are taken to be identical in both cases.

A specific feature of coherent systems is that the projection slices of a hologram are shifted radially by the value f_{po} away from the coordinate origin. Other things being equal, their resolution, defined in the first approximation by the data domain size in the frequency space, is therefore high [8].

A more important difference is that the projection $P_\vartheta(\vartheta)$ recorded by an incoherent system is a real time function. So the phase of any of the projected slices in the data domain of the frequency space is zero at the interception with the coordinate origin. The projection slices are independent of one another. In contrast, a coherent system records not only the changes in the complex envelope amplitude along the projection but also those of the phase of the echo signal carrier oscillation. As a result, in consecutive projection slices the phases of average records with the space frequency f_{po} carry information about the ranges of all unscreened scatterers of the target relative to its rotation centre. Other records of the projection slice have additional shifts with their space frequency differences with respect to the centre record. In this way, all hologram records become interrelated providing a resolution along any direction, including the transverse range.

Thus, the mathematical theory of computerised tomography for designing digital processing algorithms should be modified to adjust it to the requirements of coherent imaging. The above mathematical expressions (2.50)–(2.52) can be regarded as generalised projection slice theorems for coherent radar imaging. This enables one to employ analytical methods of computerised tomography [57] as a basis for further development of the theory of coherent imaging. Advantages of this kind of treatment are physical clarity and computation efficiency (Chapter 6).

The holographic approach to the description of inverse synthesis by coherent radars accounts for arbitrary changes in the target aspect and the frequency band width of the probing signal. Most of the available algorithms for microwave imaging have been designed for 2D viewing geometry, so digital processing for real target sizes and angle steps becomes a time-consuming endeavour. Well-elaborated mathematics of computerised tomography could considerably facilitate the development of effective computation algorithms for digital processing of 3D microwave holograms.

Chapter 3

Quasi-holographic and holographic radar imaging of point targets on the earth surface

3.1 Side-looking SAR as a quasi-holographic radar

We have shown in Chapter 2 that the aperture synthesis can be described in different ways, including a holographic approach. It was first applied by E. N. Leith to a side-looking synthetic aperture radar (SAR) [85,86]. He analysed the optical cross correlator, which processes the received and the reference signals, and concluded that 'if the reference function is a lens, the record of a point object's signal can also be considered as a lens, because the reference function has the same functional dependence as the signal itself' [85]. The signal from a point object is a Fresnel lens, and its illumination by a collimated coherent light beam creates two basic images – a real image and a virtual image (Fig. 3.1). The author also pointed out that the images formed by a Fresnel lens were identical to those created by correlation processing. He drew the conclusion that 'by reducing the optical system to only three lenses, we are, it appears, led to abolishing even these, as well as the correlation theory upon which all had been based' [85]. This was a radically new concept of SAR. The radar theory and the principles of optical processing were revised in terms of the holographic approach. Its key idea is that signal recording is not just a process of data storage, like in antenna or correlation theories, but it is rather the recording of a miniature hologram of the wave field along the carrier's trajectory. For this, the recording is made on a two-dimensional (2D) optical transparency (the 'azimuth-range'), or a complex reflected signal is recorded 2D. The first procedure uses a photographic film to record the range across the film but the azimuth and pathway range along its length. In optical recording, the image is reconstructed in the same way as in conventional off-axial holography, that is, along the carrier's pathway line. If a microwave hologram is recorded optically, its illumination by coherent light reproduces a miniature optical representation of the radar wave field. Therefore, the object's resolution is determined by the size of the hologram recorded along the pathway line, rather than by the aperture

50 *Radar imaging and holography*

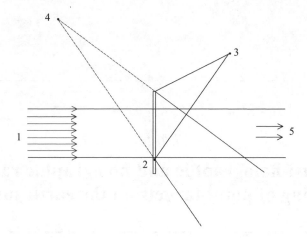

Figure 3.1 A scheme illustrating the focusing properties of a Fresnel zone plate: 1 – collimated coherent light, 2 – Fresnel zone plate, 3 – virtual image, 4 – real image and 5 – zeroth-order diffraction

of a real radar antenna. The range resolution is provided by the pulse modulation of radiated signals. Since the holographic approach to SAR is applicable only to its azimuthal channel, the authors of the work [85] termed it quasi-holographic. In his later publications on this subject, E. N. Leith pointed out that aperture synthesis should be described as a microwave analogue of holography to which holographic methods could be applied, rather than as holography proper.

Thus, a combination of SAR and a coherent optical processor represents a 'quasi-holographic' system, whose azimuthal resolution is achieved by holographic processing of the recorded wave field. Both E. N. Leith and F. L. Ingalls believe [86] that this representation is most flexible and physically clear. The use of the holographic approach for SAR analysis has so far been restricted to optical processors [87]. There is a suggestion to represent the entire SAR azimuthal channel as a holographic system [143]. In that case the initial stage of the holographic process in this channel (the formation of a microwave hologram) is the recording of the field scattered by an artificial reference source. The second stage (the image reconstruction) is described in terms of physical optics.

3.1.1 The principles of hologram recording

Let us consider a SAR borne by a carrier moving with velocity v along the x'-axis (Fig 3.2). The SAR antenna has the length L_R (the real aperture) and the beam width ϑ_R along the pathway line. The SAR irradiates the view stripe by short pulses and makes consecutive time recordings of the probing signal reflected by the object. The scattered field amplitude and phase are registered by a coherent (synchronous) detector due to the interference of the reference and received signals. This produces a multiplicative microwave hologram (Chapter 2). The role of the reference wave is played by a signal directly supplied to the synchronous detector; this is the so-called 'artificial' reference wave.

We shall describe now the receiving device of the synthetic aperture which records a hologram on a cathode tube display. Usually, a hologram is recorded by modulating the tube radiation intensity, with the photofilm moving with velocity v_f relative to the screen. For objects with different ranges R_o from the pathway line, one can use a pulse mode and vertical display scanning. As a result, the device records a series of one-dimensional (1D) holograms having different positions along the film width, depending on the distance to the respective objects. Suppose all the objects are located at a distance R_o to the pathway line. For simplicity, the radiated signal can then be taken to be continuous because the pulsed nature of the radiation is important only for the analysis of range resolution. Figure 3.3 shows an equivalent scheme of 1D microwave hologram recording. A synthetic aperture is located at point Q with the coordinates $(x', 0)$ ($x' = vt$, where t is the current moment of time), and a hypothetical source of the reference wave is at point $R(x_r, z_r)$. The source functions in a way similar to that of the reference wave during the hologram recording (Fig. 1.2). The point $P(x_o, z_o = -R_o)$ belongs to the object being viewed along the x_o-axis. If

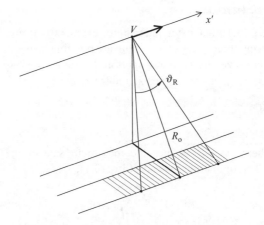

Figure 3.2   The basic geometrical relations in SAR

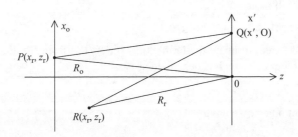

Figure 3.3   An equivalent scheme of 1D microwave hologram recording by SAR

the object's scattering characteristics are described by the function $F(x_o)$ and its size is small as compared with R_o, one can use the well-known Fresnel approximation to define the diffraction field along the $'$-axis [103]:

$$U_o(x') = C_o \frac{e^{ik_1 R_o}}{\sqrt{\lambda_1 R_o}} \int_{-\infty}^{\infty} F(x_o) e^{ik_1((x_o - x')/(R_o))} \, dx_o, \qquad (3.1)$$

where $k_1 = 2\pi/\lambda_1$ is the wave number and C_o is a complex-valued constant. The complex amplitude of the reference wave is

$$U_r(x') = A_r e^{i\varphi_r}.$$

Normally, this is a plane wave, i.e. $\varphi_r = k_1 \sin(\vartheta x')$, where ϑ is the wave 'incidence' on the hologram. The inclination of the reference wave is equivalent to that of the reference signal with a linear phase shift, providing the introduction of the carrier frequency $\omega_x = k_1 \sin \vartheta$. A coherent registration gives a hologram described as

$$h(x') = \mathrm{Re}(U_r^*(x') U_o(x')) \qquad (3.2)$$

or

$$h(x') = \mathrm{Im}(U_r^*(x') U_o(x')).$$

It follows from Eq. (3.1) that a synthetic aperture generally forms 1D Fresnel holograms. The following three types of hologram are possible, depending on the relation between the object's size, the synthetic aperture length $L_s = vT$ (T is the recording time or the time of the aperture synthesis) and the range R_o.

1. If the condition

    $$R_o \gg k_1 (x_o^2)_{\max}/2$$

    holds true (here $(x_o)_{\max}$ defines the maximum size of the object), we get Fraunhofer's approximation instead of Eq. (3.1):

    $$u_o(x') = C_o \frac{e^{ik_1 R_o}}{\sqrt{\lambda_1 R_o}} e^{ik_1((x')^2/(R_o))} \int_{-\infty}^{\infty} F(x_o) e^{-ik_1((2x' - x_o)/(R_o))} \, dx_o. \qquad (3.3)$$

    The hologram we obtain is of the Fraunhofer type.

2. If the condition

    $$R_o \gg k_1 (x_o')_{\max}^2/2 = k_1 L_S^2/8$$

    is valid, we can eliminate the term $\exp[ik_1(x')^2/R_o]$ from Eq. (3.3) to obtain a Fourier hologram, which is described as

    $$L_S \leq 2\sqrt{\lambda_1 R_o/\pi}. \qquad (3.4)$$

3. For a point object, we have

    $$F(x_o) \sim \delta(x' - x_o)$$

    and Fraunhofer's condition for diffraction becomes immediately fulfilled.

Using the filtering properties of the δ-function and Eq. (3.3), we arrive at the following equation for the hologram (with the constant phase terms ignored):

$$h(x') = A_r A_o \cos\left(\omega_x x' - k_1 \frac{(x')^2}{R_o} + 2k_1 \frac{x' x_o}{R_o}\right), \quad (3.5)$$

where A_o is the scattered wave amplitude at the receiver input.

If Eq. (3.4) holds, expression (3.5) yields

$$h(x') = A_r A_o \cos\left(\omega_x x' + 2k_1 \frac{x' x_o}{R_o}\right). \quad (3.6)$$

Thus, a synthetic aperture forms either a Fraunhofer or a Fourier hologram of a point object. The former looks like a 1D Fresnel zone plate, in accordance with Eq. (3.5), and the latter is a 1D diffraction grating with a constant step, in accordance with Eq. (3.6).

During the photographic recording, the holograms are scaled by substituting the x'-coordinate by the x-coordinate, where $x = x'/n_x$ and $n_x = v/v_f$. A constant term h_o ('displacement') is added to Eqs (3.5) and (3.6) for the photographic registration of the bipolar function $h(x')$.

3.1.2 Image reconstruction from a microwave hologram

It is reasonable to discuss the next step in the holographic process in terms of physical optics. Illumination of a photographic transparency by a plane coherent wave with the wave number k_2 produces a diffraction field, whose distribution at distance ρ from the hologram is described by the Huygens–Fresnel integral:

$$V(\xi) = \frac{e^{i(k_2\rho - \pi/4)}}{\sqrt{\lambda_2 \rho}} \int_{-v_n T/2}^{v_n T/2} h(x) e^{i(k_2/2\rho)(x-\xi)^2} dx. \quad (3.7)$$

The substitution of Eq. (3.5) into Eq. (3.7) gives

$$V(\xi) = V_o(\xi) + V_1(\xi) + V_2(\xi),$$

where $V_o(\xi)$ is the zeroth order corresponding to the displacement h_o, $V_1(\xi)$ and $V_2(\xi)$ are the functions of the reconstructed images of a point object. These functions are equal to

$$\left.\begin{array}{c}V_1(\xi)\\V_2(\xi)\end{array}\right\} = C_o \int_{-v_f T/2}^{v_f T/2} e^{i((k_2/2\rho) \pm (k_1 n_x^2/R_o))x^2} e^{-i[(k_2\xi/\rho)\pm(\omega_x n_x + (2k_1 n_x x_o/R_o))]x} dx. \quad (3.8)$$

The positions of the images along the z-axis can be found from the condition for the zeroth power of the first exponent in Eq. (3.8):

$$\rho = \pm \frac{\lambda_1 R_o}{2\lambda_2 n_x^2}. \quad (3.9)$$

Obviously, one image is virtual and the other real.

By integrating Eq. (3.8) with the condition of Eq. (3.9), we obtain

$$\left.\begin{array}{c}V_1(\xi)\\V_2(\xi)\end{array}\right\} = C_o \frac{\sin\{[\omega_x n_x + (2k_1 n_x^2/R_o)((x_o/n_x) - \xi)]v_n T/2\}}{[\omega_x n_x + (2k_1 n_x^2/R_o)((x_o/n_x) - \xi)]v_n T/2}. \tag{3.10}$$

Therefore, the image of a point object is described by the $\sin v/v$-type of function. It follows from Eq. (3.10) that the image position along the x-axis is defined by the zeroth value of the argument v, or

$$\xi = x_o/n_x + \omega_x R_o/2k_1 n_x. \tag{3.11}$$

The first term in Eq. (3.11) corresponds to the real coordinate of the object and the second one describes the carrier frequency. Images of two-point objects having the same coordinates $x_o(x_{o1} = x_{o2})$ but different ranges R_1 and $R_2(R_1 \neq R_2)$ are characterised by different coordinates ξ_1 and $\xi_2(\xi_1 \neq \xi_2)$. Therefore, the use of the carrier frequency leads to geometrical distortions of the coordinates of point objects. We should recall that the use of the carrier frequency in the first generation of SARs (with an optical processor) was necessitated by the application of Leith's off-axial holography in order to separate images from the zeroth order. The carrier frequency becomes, however, unnecessary in digital image reconstruction from complex holograms (Chapter 2).

Let us now discuss the SAR resolving power. According to Reighley's criterion, two points are thought to be separated if the major maximum of one of the $\sin x/x$ functions coincides with the first zero of the other function. This gives us the resolving power

$$\Delta x' = x_1 - x_2 = \pi R_o/k_1 L_S. \tag{3.12}$$

The aperture creating a Fraunhofer hologram with Eq. (3.5) was termed a 'focused aperture' in classical SAR theory. The focusing here is treated as a compensation for the quadratic phase shift in Eq. (3.5) during image reconstruction, the compensation being made with the transform in Eq. (3.7).

The case of an 'unfocused aperture' is described by Eq. (3.6) for the Fourier hologram, and the processing is performed with the Fourier transform of the hologram function:

$$V(\xi) = C_o \int_{-v_f T/2}^{v_f T/2} h(x) e^{-ik_2 x \xi/\rho} \, dx. \tag{3.13}$$

Eqs (3.13) and (3.6) yield

$$\left.\begin{array}{c}V_1(\xi)\\V_2(\xi)\end{array}\right\} = C_o \frac{\sin\{[(k_2/\rho)\xi \pm (\omega_x n_x + (2k_1/R_o)n_x x_o)] v_f T/2\}}{[(k_2/\rho)\xi \pm (\omega_x n_x + (2k_1/R_o)n_x x_o)] v_f T/2}. \tag{3.14}$$

Here, ρ can be taken to be the focal length of the Fourier lens.

The image position for a point object is defined as

$$\xi = \pm \frac{\rho}{k_2}\left(\omega_x n_x + \frac{2k_1}{R_o n_x x_o}\right). \tag{3.15}$$

Images of two-point objects with the same coordinates x_o but different ranges R_1 and R_2 will also be distorted due to the dependence of ξ on R_o. The resolving power from Reighley's criterion is

$$\Delta x' = x_1 - x_2 = \pi R / k_1 n_x v_f T. \tag{3.16}$$

With Eq. (3.4), the permissible limit for this parameter in SAR with unfocused processing has the value

$$\Delta x' = \sqrt{\pi \lambda_1 R_o}/4 \approx 0.44\sqrt{\lambda_1 R_o}. \tag{3.17}$$

Note that a hologram is written on a photofilm (in the case of an optical processor) or in a memory device (in the case of digital recording) continuously during the flight. For this reason, the focused or unfocused aperture regime is prescribed only at the reconstruction stage.

Synthetic aperture radar can also be considered in terms of geometrical optics, which implies phase structure analysis of a hologram. One of the expressions in (3.2) can be re-written as

$$h(x') = A_r A_o \cos(\varphi_r - \varphi_o),$$

where φ_o is the phase of the field scattered by the object. For a point object located at point P (Fig. 3.3), we can write two expressions taking into account the SAR wave propagation to the object and back:

$$\varphi_o = -2k_1(PQ - PO),$$

$$\varphi_r = -2k_1(RQ - RO),$$

where $RO = R_r$ is the distance between a hypothetical reference wave source and the coordinates origin. By expanding φ_r and φ_o into series, we get for the first-order terms

$$\varphi_r - \varphi_o \cong -\frac{4\pi}{\lambda_1}\left[(x')^2\left(\frac{1}{2R_r} - \frac{1}{2R_o}\right) - x'\left(\frac{x_r}{R_r} - \frac{x_o}{R_o}\right)\right]. \tag{3.18}$$

In a simple case of $x_o = 0$, $x_r = 0$ and $R_r = \infty$ (a plane reference wave without linear phase shift), we have

$$\varphi_r - \varphi_o = 4\pi(x')^2/2\lambda_1 R_o.$$

The space frequency in the interference pattern is

$$\nu(x) = \frac{1}{2\pi}\frac{\partial(\varphi_r - \varphi_o)}{\partial x'} = 2x'/\lambda_1 R_o. \tag{3.19}$$

At a certain value of $x'_{cr} = (L_S)_{max}/2$, the frequency ν may exceed the resolving power of the field recorder, which is defined in this case by the real aperture angle and is equal to $\nu_{cr} = 1/L_R$. From this we have the condition

$$(L_S) \leq \lambda_1 R_o / L_R = \vartheta_R R_o. \tag{3.20}$$

The substitution of $(L_s)_{max}$ into Eq. (3.12) gives a classical relation for the attainable limit of SAR resolution:

$$\Delta x_{lim} = L_R/2.$$

The pulsed nature of the signal allows determination of such an important radar parameter as the minimum repetition frequency of probing pulses, χ_{min}. Obviously, the pulse mode is similar to hologram discretisation. The distance between two adjacent records $\Delta x' = v_f/\chi$ must meet the condition

$$\Delta x' \leq [2\nu(x'_{cr})]^{-1}.$$

This condition and Eq. (3.19) gives

$$\chi_{min} = 2\vartheta/L_R.$$

By following the method suggested in Reference 92 we can obtain relations for the phase deviation of the reconstructed wave from the spherical shape (third-order wave aberrations):

$$\Delta\varphi^{(3)} = -\frac{k_2}{2}\left(D_o\frac{x^4}{4} - D_1 x^3 + D_2 x^2\right), \quad (3.21)$$

where

$$D_k = \frac{x_c^k}{R_c^3} - \frac{x_I^k}{R_I^3} \pm \frac{2\mu}{m^{4-k}}\left(\frac{x_o^k}{R_o^3} - \frac{x_{rI}^k}{R_r^3}\right),$$

x_c and R_c are the coordinates of the reconstructing wave source, $\mu = \lambda_2/\lambda_1$, $m = n_x^{-1}$.

The image coordinates for a point object are

$$\frac{1}{R_I} = \frac{1}{R_C} \pm \frac{2\mu}{m^2}\left(\frac{1}{R_o} - \frac{1}{R_r}\right),$$

$$\frac{x_I}{R_I} = \frac{x_I}{R_C} \pm \frac{2\mu}{m}\left(\frac{x_o}{R_o} - \frac{x_r}{R_r}\right).$$

The value $k = 0$ is for the spherical aberration, $k = 1$ is for the coma, $k = 2$ is for the astigmatism. These relations can be used to find the maximum size of the synthetic aperture, $(L_s)_{max}$, from Reighley's formula (wave aberrations at the hologram edges should not be larger than $\lambda_2/4$). Since spherical aberration is largest in the order of magnitude, we obtain

$$(L_S)_{max} = 2\sqrt[4]{\lambda_1 R_o^3 \Big/ \left(1 - 4\frac{\mu^2}{m^2}\right)}. \quad (3.22)$$

For typical conditions of SAR performance, the value of $(L_s)_{max}$ calculated from Eq. (3.20) is smaller than $(L_s)_{max}$ found in Eq. (3.22), that is, the effect of wave aberrations is inessential.

3.1.3 Effects of carrier track instabilities and object's motion on image quality

The carrier's trajectory instabilities are a major factor that can distort SAR images. The use of geometrical optics in the holographic approach provides a fairly simple estimation of permissible trajectory deviations from a straight line. The object's wave phase $\varphi_0(x')$ can be written as

$$\varphi_0(x') = -2k_1\left\{\left[(z_0 - g)^2 + (x' - x_0)^2\right]^{1/2} - R_0\right\},$$

where $g = g(x')$ is the trajectory deviation from the x'-axis. At $R_0 \gg x_0, x'$ and g, the binomial expansion ignoring all terms of the g^2 order gives an approximate expression for $\varphi_0(x')$:

$$\varphi_0(x') \cong -\frac{4\pi}{\lambda_1}\left(\frac{(x')^2 - 2x_0 x'}{2R_0} - \frac{x^4 - 4x_0(x')^3 + 4x_0^2(x'_0)^2}{8R_0^3}\right.$$
$$\left. -\frac{z_0 g}{R_0} + \frac{z_0 g(x')^2 - 2z_0 g x_0 x'}{2R_0^3}\right).$$

The phase equation for a wave reconstructing one of the images has a standard form:

$$\varphi_I = \varphi_c \pm (\varphi_0 - \varphi_r), \tag{3.23}$$

where φ_c are the reconstructed wave phases.

On the other hand, φ_I can be written as

$$\varphi_I = -\frac{2\pi}{\lambda_2}\left(\frac{x^2 - 2x_I x}{2R_I} - \frac{x^4 - 4x_I x^3 + 4x_I^2 x^2}{8R_I^3}\right). \tag{3.24}$$

The phases φ_c and φ_r are described by expressions similar to (3.24). The phase differences between the respective third-order terms relative to $1/R_I$ in Eqs (3.23) and (3.24) represent aberrations described as

$$\Delta\Phi = \Delta\varphi^{(3)} + \Delta\varphi_n^{(3)}.$$

The aberrations $\Delta\varphi^{(3)}$ are defined by Eq. (3.21), and $\Delta\varphi_n^{(3)}$ has the form:

$$\Delta\varphi_n^{(3)} = -k_2(D_3 g + D_4 gx - D_5 gx^2), \tag{3.25}$$

where

$$D_3 = \mp 2\mu z_0/R_0, \qquad D_4 = \mp 2\mu z_0 x_0/mR_0^3,$$
$$D_5 = \mp \mu z_0/m^2 R_0^3, \qquad m = 1/n_x, \quad \mu = \lambda_2/\lambda_1$$

and g is the trajectory deviation. Here the quantities D_3, D_4 and D_5 are aberrations arising from the trajectory instabilities.

Equation (3.25) describing distortions in the hologram phase structure can be used to calculate the compensating phase shift directly during the synthesis. For this, SAR should be equipped with a digital signal processor.

By applying Reighley's criterion to each term in Eq. (3.25), one can get the following conditions for maximum permissible deviations of the carrier's trajectory:

$$g_3 \leq \lambda_2/4/D_3 = \lambda_1 R_o/8Z_o = \lambda_1/8\cos\vartheta_o, \tag{3.26}$$

$$g_4 \leq \lambda_2/4/D_4/x_{\max} = \lambda_1 R_o^3/4L_S Z_o x_o, \tag{3.27}$$

$$g_5 \leq \lambda_2/4/D_5/x_{\max}^2 = \lambda_1 R_o^3/Z_o L_S^2. \tag{3.28}$$

Besides, if one knows the flight conditions and carrier's characteristics, Eqs (3.26)–(3.28) can be used to find constraints imposed on the parameter $\cos\vartheta_o$ and the maximum size of the synthetic aperture:

$$\cos\vartheta_o \leq \lambda_1/8g,$$

$$L_{S_{\max}} \leq \lambda_1 R_o^2/4gx_o,$$

$$L_{S_{\max}} \leq R_o\sqrt{\lambda_1/g}.$$

Normally, SAR meets the conditions $L_S \ll R_o$ and $x_o \ll R_o$. So D_4 and D_5 can be neglected leaving only the factor D_3, which severely restricts the trajectory stability (see Eq. (3.26)).

Effects arising in a synthetic aperture during the viewing of moving targets can be estimated in terms of physical optics. Suppose a point object moves radially (along the z-axis) at velocity v_o, such that its displacement is smaller than the range resolution for the synthesis time T. Then, the equation for the hologram, ignoring constant phase terms, is

$$h(x) \sim \cos\left(\omega_x n_x x + 2k_1\frac{v_o}{v}n_x x - k_1\frac{n_x^2 x^2}{R_o} + 2k_1\frac{n_x x_o x}{R_o} - \frac{k_1}{R_o}\left(\frac{v_o}{v}\right)^2 n_x^2 x^2\right). \tag{3.29}$$

The substitution of Eq. (3.29) into (3.7) gives a condition for viewing the focused image:

$$\rho = \pm\left(\frac{k_2}{2k_1}\frac{R_o}{n_x^2}\right)\Big/\left[1+\left(\frac{v_o}{v}\right)^2\right].$$

Since $v_o/v \ll 1$, the image can be viewed practically in the same plane as that for an immobile object. Keeping this in mind, we can obtain, after the integration, a function describing one of the reconstructed images:

$$V(\xi) = C_o \frac{\sin\{[\omega_x n_x + 2k_1(v_o/v)n_x + (2k_1 n_x/R_o)(x_o - n_x\xi)]v_f T/2\}}{[\omega_x n_x + 2k_1(v_o/v)n_x + (2k_1 n_x/R_o)(x_o - n_x\xi)]v_f T/2}.$$

The image position is defined as being

$$\xi = \frac{x_o}{n_x} + \frac{\omega_x R_o}{2k_1 n_x} + \frac{R_o}{n_x}\frac{v_o}{v}.$$

Clearly, the object's motion is equivalent to the use of additional carrier frequency at the recording stage, which causes the image shift. The optical processor deals with a real image recorded on a photofilm. The recording field on the film is limited by a diaphragm cutting off the background. The value of v_o may become so large that no image will be recorded because of the shift.

The object's motion in the azimuthal direction (along the x'-axis) at velocity v_o is equivalent to the change in the SAR's flight velocity. Then Eq. (3.9) describing the position of the focused image along the z-axis can be re-written as

$$\rho' = \pm\lambda_1 R_o/2\lambda_2 n_x^2 = \pm\lambda_1 R_o v^2/2\lambda_2(v-v_o)^2.$$

Therefore, the object's motion along the x'-axis changes the focusing conditions by the value

$$\delta\rho = \rho' - \rho = 2\rho\frac{v_o}{v}\left(1 - \frac{v_o}{2v}\right) \Big/ \left(1 - \frac{v_o}{v}\right)^2, \qquad (3.30)$$

where ρ is found from Eq. (3.9). If the condition $v_o \ll v$ is fulfilled, we have

$$\delta\rho \approx 2\rho v_o/v. \qquad (3.31)$$

Equation (3.30) yields

$$v_o = v\left(1 - \sqrt{1 - \delta\rho/(\rho + \delta\rho)}\right).$$

On the other hand, a simple geometrical consideration can give the following relations for the resolving power of SAR along the z-axis (longitudinal resolution):

$$\Delta\rho = 2(\Delta x' v_f)^2/\lambda_2 v^2 = 2(\Delta x')^2/\lambda_2 n_x^2. \qquad (3.32)$$

The focusing depth $\Delta\rho$ is defined as the focal plane shift along the z-axis by a distance, at which the azimuthal resolution $\Delta x'$ becomes twice as poor as the diffraction limit in Eq. (3.12).

The viewing of a focused image of an object moving at velocity v_o requires an additional focusing of the optical processor. The object's velocities that require the focusing can be found from the condition $\Delta\rho < \delta\rho$, where $\delta\rho$ is given by Eq. (3.31). Using Eqs (3.9) and (3.32), we get

$$v_o > 2(\Delta x')^2/\lambda_1 R_o.$$

At lower velocities, there is no need to re-focus the processor, and a poorer image quality may be assumed to be inessential.

To conclude Section 3.1, we should like to emphasise the following. The SAR operation principles can be described by conventional methods (Chapter 2) that are still widely used [73] or with a holographic approach representing the side-looking synthetic aperture and the processor as an integral system for recording and reconstructing the wave field. The analysis of the aperture synthesis can be based on the well-elaborated principles of holography as well as on physical and geometrical optics. The examples we have discussed support the physical clarity of the holographic approach and its value for SAR analysis. We can get a better insight into the

3.2 Front-looking holographic radar

The operation principle of a front-looking holographic radar was discussed in Chapter 2. A high resolution across the pathway line (Fig. 3.4) is provided in it by a multibeam antenna pattern of a large receiving antenna array located, say, along the aircraft wings [72]. The resolution along the pathway line is achieved by the aperture synthesis. There is another radar design, in which the desired transversal resolution is provided by a phased antenna array mounted under the fuselage and the longitudinal resolution by a synthetic aperture [81,82].

3.2.1 The principles of hologram recording

A coherent transmitter (Fig. 3.5) generates a continuous or pulsed signal (to decouple the transmitter and the receiver) and illuminates the desired survey zone under the aircraft. The receiving antenna represents a linear or phased array of numerous receivers. The amplitude and phase of the reflected signal are recorded by each array element for the time T_s, synthesising a 2D aperture of size $X_s Y$ along the trajectory segment $X_s = vT_s$. Signals at the receiver output are saved by a memory unit, for example, on a photofilm [81]. The film record can be regarded as a 2D plane optical hologram equivalent to a microwave hologram with the size $X_s Y$ (Fig. 3.4). If the radar has an optical processor, it reconstructs the wave front recorded on the optical hologram to produce an optical image of the earth surface within the view zone. Thus, the operation principle of this type of radar is totally holographic and it is reasonable to call

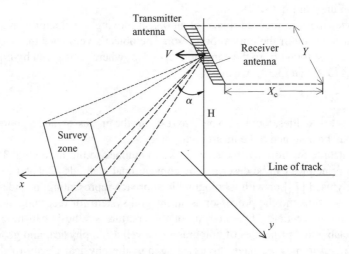

Figure 3.4 The viewing field of a holographic radar

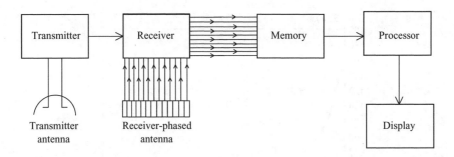

Figure 3.5 A schematic diagram of a front-looking holographic radar

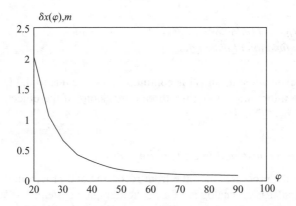

Figure 3.6 The resolution of a front-looking holographic radar along the x-axis as a function of the angle φ

it a front-looking holographic radar [72,81]. Since it is an analogue of a 2D optical holographic system, it produces a 3D image. The resolution of a holographic radar can be examined by analysing the uncertainty function [72]. The slicing of this function into equal power levels at the point of 0.7 gives an approximate radar resolution:

$$\delta y = 0.88\lambda_1 H/Y \sin\varphi, \tag{3.33}$$

$$\delta x = 0.45\lambda_1 H/X_s \sin^3\varphi, \tag{3.34}$$

$$\delta z = 7\lambda_1 H^2/(2X_s^2 \sin^3\varphi + Y^2 \sin^3\varphi), \tag{3.35}$$

where X_s is the synthetic aperture length, $\varphi = 90° - \alpha$.

Figures 3.6 and 3.7 show the dependence of δx and δz on the angle φ, plotted from the following initial parameters: $\lambda_1 = 1.78$ cm, $H = 300$ m, $Y = 1$ m and $X_s = 30$ m. One can see that a holographic radar possesses a fairly large resolving power.

It follows from Eqs (3.33), (3.34) and (3.35) that in addition to the 'conventional' resolution along and across the pathway line, a holographic radar has a longitudinal

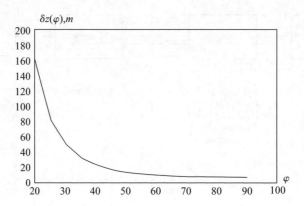

Figure 3.7 *The resolution of a front-looking holographic radar along the z-axis as a function of the angle φ*

resolution δz even when its signal is continuous. This is due to the fact that a hologram contains information about the three dimensions of the object, including the longitudinal range (Chapter 2).

3.2.2 Image reconstruction and scaling relations

Consider now the processes of wave front recording and processing in this type of radar. As the radar is an analogue of a 2D holographic system, it would be natural to analyse it in terms of the holographic approach developed in Section 3.1, which treats the radar and the processing unit as an integral system. For this, we shall examine a generalised hologram geometry [50,51]. Suppose a wave comes from a microwave point source with the coordinates (x_o, y_o, z_o), and a reference wave is generated by a point source with the coordinates (x_r, y_r, z_r), as shown in Fig. 3.8(a). The wave field being recorded has the wavelength λ_1.

At the second stage, the recorded hologram is illuminated by a spherical wave with the wavelength λ_2, coming from a point source with the coordinates (x_p, y_p, z_p), as shown in Fig. 3.8(b). A paraxial approximation will then give the coordinates of two reconstructed images:

$$\left. \begin{aligned} x_i &= \pm \frac{\lambda_2 z_i}{\lambda_1 z_o} x_o \mp \frac{\lambda_2 z_i}{\lambda_1 z_r} x_r - \frac{z_i}{z_p} x_p \\ y_i &= \pm \frac{\lambda_2 z_i}{\lambda_1 z_o} y_o \pm \frac{\lambda_2 z_i}{\lambda_1 z_r} y_r - \frac{z_i}{z} y_p \\ z_i &= \left(\frac{1}{z_p} \pm \frac{\lambda_2}{\lambda_1 z_r} \mp \frac{\lambda_2}{\lambda_1 z_o} \right)^{-1} \end{aligned} \right\} \quad (3.36)$$

The upper arithmetic signs in the equalities of (3.36) are for the virtual image and the lower ones are for the real image. When z_i is positive, the image is virtual and is on the left of the hologram; when z_i is negative, the image is real and is located on

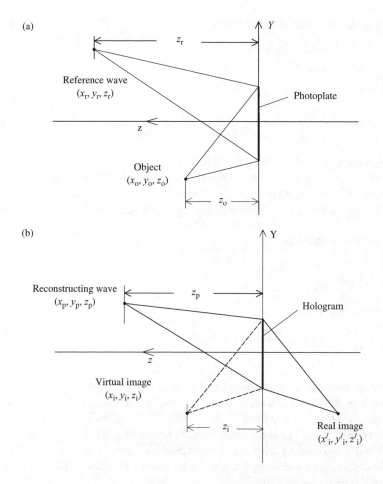

Figure 3.8 Generalised schemes of hologram recording (a) and reconstruction (b)

the right of the hologram. At $\lambda_1 = \lambda_2$, $z_r = z_o$ and $z_c > 0$ both images are virtual, whereas at $\lambda_1 = \lambda_2$, $z_r = z_o$ and $z_c < 0$ they are real.

One can show with Eqs (3.36) that holographic images of objects more complex than just a point, for example, consisting of two point sources, can be magnified or diminished relative to the respective object [50,51].

As the reconstructed wave front is 3D, the transverse (along the x- and y-axes) and the longitudinal (along the z-axis) magnifications obtained during the reconstruction can be analysed separately.

From Eq. (3.36), the transverse magnifications are:

for the real image (superscript 'r')

$$M_t^r = \frac{\partial x_i}{\partial x_o} = \frac{\partial y_i}{\partial y_o} = \frac{\lambda_2 z_i}{\lambda_1 z_o}, \qquad (3.37)$$

for the virtual image (superscript 'v')

$$M_t^v = \frac{\partial x_i}{\partial x_o} = \frac{\partial y_i}{\partial y_o} = -\frac{\lambda_2 z_i}{\lambda_1 z_o} \qquad (3.38)$$

or

$$M_t^r = M_t^v = \left|1 - \frac{z_o}{z_r} \mp \frac{\lambda_1 z_o}{\lambda_2 z_p}\right|^{-1}. \qquad (3.39)$$

Here the superscript is for the real image and the subscript is for the virtual one. The transverse magnification describes the ratio of the width and height of the image to the appropriate parameters of the real object.

The longitudinal magnification can be found by differentiating Eq. (3.36) for z_i:

for the real image

$$M_l^r = \frac{\partial z_i}{\partial z_o} = \frac{\lambda_2 z_i^2}{\lambda_1 z_o^2} \simeq \frac{\lambda_1}{\lambda_2}(M_t^r)^2 \qquad (3.40)$$

for the virtual image

$$M_l^v = \frac{\partial z_i}{\partial z_o} = -\frac{\lambda_2 z_i^2}{\lambda_1 z_o^2} \simeq -\frac{\lambda_1}{\lambda_2}(M_t^v)^2. \qquad (3.41)$$

The longitudinal magnification of a virtual image is always negative. This means that the image always has a relief inverse to that of the object: it is pseudoscopic.

Equations (3.37), (3.38) and (3.40), (3.41) show that the longitudinal and transverse magnifications are not identical, so the image of a 3D object is distorted. The matter is that the object's relief cannot be reproduced exactly in an image. The condition for obtaining an undistorted image can be derived from the equality of transverse and longitudinal magnifications:

$$M_t^r = M_l^r \quad \text{or} \quad \frac{\lambda_2 z_i}{\lambda_1 z_o} = \frac{\lambda_2 z_i^2}{\lambda_1 z_o^2}.$$

Therefore, a geometrical similarity is possible only if the image is reconstructed at the site the object occupied during the recording.

By substituting the coordinate $z_i = z_o$ into Eq. (3.36), we can get an expression for the coordinates of the reconstructing source:

$$\frac{1}{z_p} = \frac{1}{z_o}\left(1 \mp \frac{\lambda_2}{\lambda_1}\right) \mp \frac{\lambda_2}{\lambda_1 z_r}. \qquad (3.42)$$

Another way of obtaining an undistorted image is to change the scale of the linear hologram size by a factor of m at the transition from the recording to the reconstruction [50]. At $m < 1$, the hologram becomes smaller while at $m > 1$ it becomes larger. The coordinates of an image reconstructed from a hologram diminished m times can be found from

$$\begin{aligned}
x_i &= \pm m\frac{\lambda_2 z_i}{\lambda_1 z_o}x_o \mp m\frac{\lambda_2 z_i}{\lambda_1 z_r}x_r - \frac{z_i}{z_p}x_p; \\
y_i &= \pm m\frac{\lambda_2 z_i}{\lambda_1 z_o}y_o \mp m\frac{\lambda_2 z_i}{\lambda_1 z_r}y_r - \frac{z_i}{z_p}y_p; \\
z_i &= \left(\frac{1}{z_p} \pm m^2\frac{\lambda_2}{\lambda_1 z_r} \mp m\frac{\lambda_2}{\lambda_1 z_o}\right)^{-1}.
\end{aligned} \quad (3.43)$$

The transverse magnifications are:

for the real image (superscript 'r')

$$M_t^r = \frac{\partial x_i}{\partial x_o} = \frac{\partial y_i}{\partial x_o} = m\frac{\lambda_2 z_i}{\lambda_1 z_o} \quad (3.44)$$

for the virtual image (superscript 'v')

$$M_t^v = \frac{\partial x_i}{\partial y_o} = \frac{\partial y_i}{\partial y_o} = -m\frac{\lambda_2 z_i}{\lambda_1 z_o}. \quad (3.45)$$

The longitudinal magnifications are:

for the real image

$$M_l^r = \frac{\partial z_i}{\partial z_o} = m^2\frac{\lambda_1}{\lambda_2}(M_t^r)^2, \quad (3.46)$$

for the virtual image

$$M_l^v = \frac{\partial z_i}{\partial z_o} = -m^2\frac{\lambda_1}{\lambda_2}(M_t^v)^2. \quad (3.47)$$

The condition for obtaining an undistorted image also follows from the equality of transverse and longitudinal magnifications

$$z_o = mz_i. \quad (3.48)$$

The substitution of Eq. (3.48) into Eq. (3.36) yields the coordinates of the reconstructing source; in particular, for z_p we have

$$\frac{1}{z_p} = \frac{m}{z_o}\left(1 \pm m^2\frac{\lambda_2}{\lambda_1}\right) \mp m\frac{\lambda_2}{\lambda_1 z_r}. \quad (3.49)$$

If the recorded hologram is magnified m times, the reconstructed image is at a distance

$$z_i = \left(\frac{1}{z_p} \pm \frac{\lambda_2}{\lambda_1 z_r m^2} \mp \frac{\lambda_2}{\lambda_1 z_o m^2}\right)^{-1}, \quad (3.50)$$

from this hologram, and the transverse magnifications are:

for the real image

$$M_t^r = \frac{\lambda_2 z_i}{\lambda_1 z_o m} \tag{3.51}$$

for the virtual image

$$M_t^v = -\frac{\lambda_2 z_i}{\lambda_1 z_o m}. \tag{3.52}$$

In the case of imaging a 3D object, the distortions due to the difference in the transversal and longitudinal directions will be minimal at $M_t = \lambda_2/\lambda_1$. Then Eqs (3.51) and (3.52) give $M_t = M_l$. The distortions of the real and virtual images due to the shift are also eliminated at $a = b = 0$ (Fig. 3.9), but the images and the zeroth-order overlap, a situation unacceptable for optical holography. In a holographic radar capable of recording a complex hologram (Chapter 2), there is no problem with decoupling a single image and the zeroth order.

We shall now turn to the limiting longitudinal resolution in a holographic radar and consider the recording and reconstruction schemes (Fig. 3.9 (a) and (b), respectively) in order to define longitudinal magnifications. Using a paraxial approximation, the authors of the work [142] have shown that the minimal resolvable longitudinal distance for a reconstructed real image is written as

$$(d_r)_{\min} \cong \Delta l_r \quad \text{at} \quad d \ll R_1, \tag{3.53}$$

where

$$\Delta l_r = l_r' - l_r'',$$
$$l_r' = \lambda_1 R_1 L_1 L_2 / (\lambda' L_1 L_2 - \lambda' R_1 L_2 - \lambda_1 R_1 L_1), \tag{3.54}$$
$$l_r'' = \lambda_1 R_1 L_1 L_2 / (\lambda'' L_1 L_2 - \lambda'' R_1 L_2 - \lambda_1 R_1 L_1), \tag{3.55}$$

λ' and λ'' are the minimal and maximal wavelengths of the reconstructing beam. If the distance d is small compared to R_1, the longitudinal magnification is

$$M_l^r = \frac{d_r}{d} \quad \text{at} \quad d \ll R_1.$$

Hence, we have

$$(d_r)_{\min} \geq \Delta l_r (M_l^r)^{-1}, \tag{3.56}$$

where

$$M_l^r = \lambda_1 \lambda_2 (L_1 L_2)^2 / [\lambda_2 L_1 L_2 - \lambda_2 R L_2 - \lambda_1 R_1 L_1]^2$$

and $\lambda_2 = (\lambda' \lambda'')^{1/2}$ is the average wavelength of the reconstructing source. Similar expressions can be derived for the reconstruction of a virtual image.

The analysis we have made allows one to choose suitable recording and reconstruction procedures when one uses a holographic radar. Clearly, the parameters of these procedures are closely interrelated, so the radar and its processor should be regarded as an integral system.

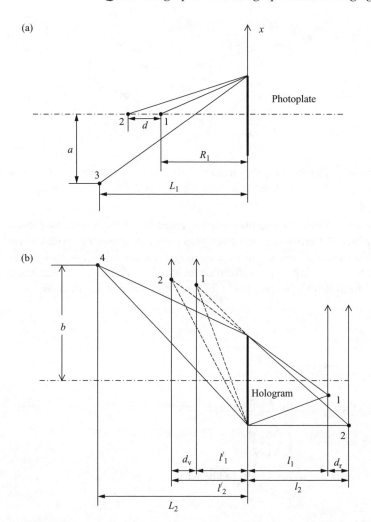

Figure 3.9 Recording (a) and reconstruction (b) of a two-point object for finding longitudinal magnifications: 1, 2 – point objects, 3 – reference wave source and 4 – reconstructing wave source

3.2.3 The focal depth

Consider now the focal depth of an image produced by a holographic radar. Chapter 2 discussed the problem of recording a 3D image in a 2D medium using a classical holographic approach. The image quality then depends on the focal depth of the image. The process of reconstruction gives the opportunity to obtain a 3D image of a scene. Following the reconstruction, the image is again recorded in a 2D medium, so the problem of focal depth arises once more. This parameter can be defined by analogy with the recommendations suggested in Reference 96.

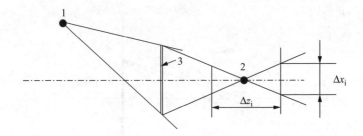

Figure 3.10   The focal depth of a microwave image: 1 – reconstructing wave source, 2 – real image of a point object and 3 – microwave hologram

The focal depth of a microwave holographic image is a longitudinal distance Δz_i, along which the cross section of the beam reconstructing the virtual or real image of a point object is smaller than the resolution elements Δx_i, so that it is perceived as a point image (Fig 3.10). A formula for the focal depth of a virtual image can be derived from Eqs (3.40) and (3.41) for the longitudinal magnification:

$$\Delta z_i = M_l^v \Delta z_o \tag{3.57}$$

or

$$\Delta z_i = -\frac{\lambda_1}{\lambda_2}(M_t^v)^2 \Delta z_o. \tag{3.58}$$

With the relation for the transverse magnification (3.52), one can write

$$\Delta z_i = -\frac{\lambda_1}{\lambda_2} \Delta x_i \left(\frac{\Delta z_o}{\Delta x_o} \right). \tag{3.59}$$

The last factor in Eq. (3.59) can be written as

$$\frac{\Delta z_o}{\Delta x_o} = tg\alpha_o, \tag{3.60}$$

where α_o is the aperture angle in the objects' space. Then Eqs (3.59) and (3.60) yield

$$\Delta z_i = -\frac{\lambda_1}{\lambda_2} \left(M_t^v\right) \frac{\Delta x_i}{tg\alpha_o}. \tag{3.61}$$

If the scale of the initial hologram is diminished m times, we have

$$\Delta z_i = -m^2 \frac{\lambda_1}{\lambda_2} \left(M_t^v\right) \frac{\Delta x_i}{tg\alpha_o}. \tag{3.62}$$

Let us now define the quantity Δx_i. Although the resolution along the x- and y-axes is determined by different physical conditions, the resolution elements Δx and Δy must have the same values. Therefore, instead of Δx_i one can use δx describing the resolution along the pathway line provided by the aperture synthesis. Then Eqs (3.62) and (3.34) give

$$\Delta z_i = -\frac{0.45\lambda_1^2 M_t^v H}{\lambda_2 X_s tg\alpha_o \sin^3 \varphi}. \tag{3.63}$$

A characteristic feature of this expression is that Δz_i is inversely proportional to the synthetic aperture length X_s.

It is also worth discussing some practical aspects of scaling in a holographic radar. Unlike SAR, this type of radar has no anamorphism, that is, the image planes coincide in azimuth and range. So there is no need to use special optics to eliminate anamorphism. However, the image proportions along the x- and y-axes do not coincide because the scaling coefficient in azimuth, P_x, differs from that in range, P_y. According to Reference 81, P_x is defined as

$$P_x = v/V, \tag{3.64}$$

where v is the velocity of the transparency on which the hologram is recorded and V is the velocity of the antenna array.

Along the y-axis, the scaling coefficient P_y is

$$P_y = \frac{W}{2a}, \tag{3.65}$$

where W is the transparency width and $2a$ is the double length of the antenna array. As a result, the holographic image appears to be defocused along the x- and y-axes. The image scale along these axes can be equalised by special optics – spherical or cylindrical telescopes. The optics suggested in Reference 81 can change the image scale from 4 to 25 times. Transversal and longitudinal scales of an image can be equalised by choosing a proper coefficient m. Therefore, the final values of longitudinal magnification and focal depth can be found only after one has selected all the scaling coefficients P_y, P_x and m.

To conclude, we summarise specific features of front-looking SAR systems.

1. It has been shown in Reference 74 that SAR systems have a serious limitation. When the view zone approaches the pathway line, the resolution in azimuth becomes much poorer. This makes it impossible to obtain quality images in the front view zone. In contrast, a holographic radar provides a high resolution directly under the aircraft.
2. Another essential advantage of a holographic radar is a high longitudinal resolution even in a continuous pulse mode along the z-axis, providing 3D relief images.
3. The 3D character of a holographic radar image is a basis for obtaining range contour lines which can then be recalculated to get surface contours [81]. This operation mode is 'purely' holographic. In fact, it implements the principle of two-frequency holographic interferometry.
4. A high 3D quality of the image requires the use of a new parameter – the image focal depth, by analogy with optical systems.

5. The view field geometry in a holographic radar is equivalent to that of airborne infrared and optical devices, so it is possible to combine microwave images with infrared and optical images. This kind of complexing considerably increases the radar capability to detect and identify targets.

3.3 A tomographic approach to spotlight SAR

3.3.1 Tomographic registration of the earth area projection

Today there are two practically valuable cases when tomographic algorithms can be used for reconstruction of radar images: inverse aperture synthesis by rotating an object round its centre of mass (see Chapter 6) and aperture synthesis in a spotlight or telescopic mode [100]. We shall analyse the latter case.

A microwave radar with a synthetic aperture borne by a carrier and operating in a spotlight mode has a real antenna oriented onto an earth area to be surveyed. The area is illuminated for a longer time than is normally done in stripe surface mapping [100], so this type of SAR has a greater resolving power than a conventional side-looking SAR. Figure 3.11 shows the basic geometrical relations illustrating the spotlight mode. For simplicity, we shall consider a 2D case. Suppose that the coordinate origin is related to a certain point on the earth's surface; the x-axis is the range and the y-axis is the azimuth. During the carrier flight, a real antenna ray is incident onto this area at an angle ϑ to the x-axis. The SAR scans the target with wideband pulses, for example, linear frequency modulation (LFM) pulses of the Re$S(t)$ type, where

$$S(t) = \begin{cases} e^{j(\omega_0 t + \alpha t^2)}, & |t| \leq \tau/2, \\ 0, & \text{otherwise} \end{cases} \qquad (3.66)$$

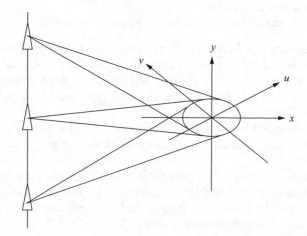

Figure 3.11   *The basic geometrical relations for a spot-light SAR*

where ω_0 is the SAR carrier frequency, 2α is the LFM slope and τ is the pulse duration. Note that the latter condition is not obligatory because the signal may have a narrow band. It is assumed that the target is in the far zone and the microwave phase front in the target vicinity is planar. The signal reflected by a unit area of the surface at the point (x_0, y_0) is

$$r_0(t) = A\text{Re}\left\{g(x_0, y_0)S(t - \frac{2R}{c})\right\} dx\, dy, \qquad (3.67)$$

where A is the amplitude coefficient accounting for the signal attenuation during the propagation; $2R/c$ is the time delay of the signal while it covers the distance R in both directions; $g(x,y) = |g(x,y)| \exp[j\varphi_0(x,y)]$ is the density function, whose physical sense here is just the distribution of the earth surface reflectivity; and $\varphi_0(x,y)$ is the signal phase shift due to the reflection. We also assume that the function $g(x,y)$ remains constant within the given ranges of radiation frequencies and view angles ϑ.

Normally, when the distance to the target is much larger than the target's size, elements of the ellipses in Fig. 3.11 may be regarded as segments of straight lines. Therefore, with Eqs (1.17) and (3.67) we can write down the total echo signal from all reflectors located within a narrow band normal to the u-axis and having the width du at $u = u_0$:

$$r_1(t) = A\text{Re}\left\{p_\vartheta(u_0)S\left[t - \frac{2(R_o + u_0)}{c}\right]\right\} du,$$

where R_o is the distance between the SAR and the target centre.

The total signal from the area being surveyed is

$$r_\vartheta(t) = A\text{Re}\left\{\int_{-L}^{L} p_\vartheta(u)S\left[t - \frac{2(R_o + u_0)}{c}\right] du\right\}, \qquad (3.68)$$

where L is the area length along the u-axis and $A = \text{const}$, which is valid at $R_o \gg L$. In contrast to the classical situation presented in Fig. 1.6, the linear integral used for the projection is taken along the line normal to the microwave propagation direction.

Now we substitute Eq. (3.66) for the LFM pulse into Eq. (3.68), simultaneously detecting the received signal with a couple of quadrature multipliers, and then we pass the output signals through low-frequency filters. What we eventually get is the signal

$$c_\vartheta(t) = \frac{A}{2}\int_{-L}^{L} p_\vartheta(u) \exp\left(j\frac{4\alpha u^2}{c^2}\right) \exp\left\{-j\frac{2}{c}[\omega_0 + 2\alpha(t - \tau_0)u]\right\} du,$$

where

$$\tau_0 = 2R_o/c \quad \text{and} \quad \tau/2 + 2(R_o + L)/c \leq t \leq \tau/2 + 2(R_o - L)/c. \qquad (3.69)$$

The latter expression is the Fourier transform of the function $p_\vartheta(u) \exp(j4\alpha u^2/c^2)$, whose exponential factor can be easily eliminated if we find the inverse Fourier transform of $c_\vartheta(t)$ by multiplying the result by $\exp(-j4\alpha u^2/c^2)$ and making Fourier transform again. This quadratic phase factor can quite often be neglected. Eventually, we have

$$c_\vartheta(t) = \frac{A}{2} P_\vartheta \left\{ \frac{2}{c}[\omega_0 + 2\alpha(t - \tau_0)] \right\}, \tag{3.70}$$

where the time t satisfies Eq. (3.69).

Therefore, if one uses LFM pulses, demodulated signals received from every illuminated direction are part of a 1D Fourier function of the centre projection of the earth area at the respective view angle. In other words, the processor output signal represents a Fourier image of the projection function (within the time interval considered) and the data are registered in Fourier space. In accordance with the projection theorem, the function (3.70) is a cross section, taken at the angle ϑ, of the 2D Fourier transform $G(X, Y)$ of the desired density function $g(x, y)$. It follows from Eq. (3.69) that the function $P_\vartheta(X)$ is defined in the range $X_1 \leq X \leq X_2$, where

$$X_1 = \frac{2}{c}\left(\omega_0 - \alpha\tau + \frac{4\alpha L}{c}\right) \cong \frac{2}{c}(\omega_0 - \alpha\tau),$$

$$X_2 = \frac{2}{c}\left(\omega_0 + \alpha\tau - \frac{4\alpha L}{c}\right) \cong \frac{2}{c}(\omega_0 + \alpha\tau). \tag{3.71}$$

Since measurements are usually limited to a certain range of angles $\vartheta_{\min} \leq \vartheta \leq \vartheta_{\max}$, it is clear that the counts of $G(X, Y)$ can be obtained at the polar grid points within a limited circular segment (the shaded region in Fig. 1.7). The inner and outer radius of the circle, X_1 and X_2, are proportional to the smallest ($\omega_0 - \alpha\tau$) and the largest ($\omega_0 + \alpha\tau$) frequency values of the LFM pulse.

Further, one can employ classical algorithms based on interpolations and inverse Fourier transforms to reconstruct $g(x, y)$. Before performing the latter procedure, it is useful to multiply the G function by the weight or 'window' function, to reduce parasitic side lobes in the image.

3.3.2 Tomographic algorithms for image reconstruction

The next step in developing this algorithm is to perform a 2D inverse Fourier transform in the polar coordinates. This can be done as follows. First, introduce the function

$$F(x, y) = \sum_{(u,v)\in P} G(u, v)\delta(x - u, y - v),$$

where P is a polar grid; $\delta(\cdot)$ is the delta-function of Dirac; $G = S \cdot W$, where S is a complex-valued function prescribed by P (experimental data); and W is a real-valued weight function. We should also prescribe the real parameters $a > 0, b > 0$ and the integer parameters $M > 0$ and $N > 0$ such that the rectangular grid

$$R = \{(ma, nb) | -M/2 \leq m < M/2, -N/2 \leq n < N/2\},$$

should satisfy a discrete representation of the object sought for.

The quantity $M \times N$ is equal to the number of pixels on the image, each pixel having the size $a \times b$. According to the sampling theorem, I/a and I/b are approximately equal to the size of the P region along the x- and y-axes, while $I/(Ma)$ and $I/(Nb)$ should equal the spacing between the grid nodes along the same axes. Thus, the P grid consists approximately of M radial lines and N pixels along each line. Note that in classical tomography, we have $M \cong N$ and the grid P includes about $\pi M/2$ radial lines and N pixels along each line.

We can now estimate the inverse Fourier transform f of the function F across the region R:

$$f(ma, nb) = \iint F(x,y) E(xma + ynb) \, dx \, dy$$
$$= \sum_{(u,v) \in P} G(u,v) E(uma + vnb)$$

with $E(z) = \exp(j2\pi z)$.

A straightforward calculation of exact values of f in the region R using the last formula will require about $M^2 N^2$ elementary arithmetic operations. For f estimations in this region, however, one can employ conventional methods with a smaller number of operations, using the interpolation algorithm mentioned above and the convolution algorithm to be discussed below. They are a fairly simple algorithm for a rigorous calculation of the functions $f(ma, nb)$ with the so-called homogeneous concentrically square polar grid, which requires about $MN \log_2(MN)$ operations.

The polar grid P is described as

$$P = \{(u(i,k) = A(k) + iB(k), v(i,k) = C + kD)$$
$$\text{at} \quad 0 \leq i < M, \ 0 \leq k < N\},$$

where $A(k) = -(C + kD) tg(\vartheta_o/2)$, $B(k) = -2A(k)/(M-1)$, ϑ_o is the size of the R region, C and D are some selected real positive numbers.

The values of the f function are found in two steps. First, for $-M/2 \leq m < M/2$ and $0 \leq k < N$ we find the function

$$H(m,k) = E(m^2 a B(k)/2) \sum_{i=0}^{i=M-1} \{G(i,k) E(i^2 a B(k)/2)\}$$
$$\times E(-(m-i)^2 a B(k)/2).$$

Second, for $-M/2 \leq m < M/2$ and $-N/2 \leq n < N$ we calculate the desired function

$$f(ma, nb) = E(nbC) \sum_{k=o}^{N-1} \{H(m,k) E(maA(k))\} E(nk/N).$$

Consider now a tomographic algorithm for reconstruction of SAR images, based on the convolution back projection (CBP) method. It employs the relation between the functions $g(x, y)$ and $G(X, Y)$ written in the polar coordinates [34]:

$$g(\rho \cos \Phi, \rho \sin \Phi) = \frac{1}{4\pi^2} \int\limits_{-\pi/2}^{\pi/2} \int\limits_{-\infty}^{\infty} G(r \cos \vartheta, r \sin \vartheta)|r|$$

$$\times \exp[jr\rho \cos(\Phi - \vartheta)]\, dr\, d\vartheta.$$

With the projection theorem, the last expression can be re-written as

$$g(\rho \cos \Phi, \rho \sin \Phi) = \frac{1}{4\pi^2} \int\limits_{-\pi/2}^{\pi/2} \int\limits_{-\infty}^{\infty} P_\vartheta(r)|r| \exp[jr\rho \cos(\Phi - \vartheta)]\, dr\, d\vartheta. \quad (3.72)$$

The integral around the variable r can be interpreted as inverse Fourier transform with the argument $\rho \cos(\Phi - \vartheta)$; from the convolution theorem, Eq. (3.72) reduces to

$$g(\rho \cos \Phi, \rho \sin \Phi) = \frac{1}{2\pi} \int\limits_{-\pi/2}^{\pi/2} (P_\vartheta * k_\mathrm{r})\rho \cos(\Phi - \vartheta)\, d\vartheta, \quad (3.73)$$

where k_r is the Fourier transform of the function $|r|$.

The algorithm used in computer-aided tomography (CAT) involves the calculation of the $P_\vartheta * k_\mathrm{r}$ convolution for each value of ϑ, followed by an approximate integration around the variable ϑ by summing up the results obtained. Since one measures the function $P_\vartheta(r)$, the reconstruction algorithm should be based on Eq. (3.72) rather than Eq. (3.73). It follows from Eq. (3.72) that $P_\vartheta(r)$ must be known for all r values, but it is clear from the foregoing (see Eq. (3.71)) that $P_\vartheta(r)$ is known only for a limited range of r values with the centre at $r = 2\omega_0/c$. Besides, the circular segment of the P_ϑ function (Fig. 3.12) should be shifted towards the origin. With these remarks in mind, Eq. (3.72) can be reduced to

$$g(\rho \cos \Phi, \rho \sin \Phi) = \frac{1}{4\pi^2} \int\limits_{\vartheta_{\min}}^{\vartheta_{\max}} \left\{ \int\limits_{0}^{X_2 - X_1} P_\vartheta(r + X_1)|r + X_1|W_1(r) \right.$$

$$\left. \times \exp[jr\rho \cos(\vartheta - \Phi)]\, dr \right\} \times W_2(\vartheta) \exp[jX_1\rho \cos(\Phi - \vartheta)]\, d\vartheta. \quad (3.74)$$

where $W_1(r)$ and $W_2(\vartheta)$ are additional weight functions [33].

The interpolation and convolution algorithms have been compared quantitatively. The comparison is based on two criteria: (1) the level of multiplicative noise (side lobes)

$$R_{MN} = 10 \lg \left| \frac{N_{\mathrm{oml}}^2}{N_{\mathrm{iml}}^2} \right|,$$

where N_{oml} is the number of pixels outside the major lobe on a point scatterer's image and N_{iml} is the number of pixels inside the major lobe; and (2) the computation time and complexity, or the number of elementary arithmetic operations to be made. The value of R_{MN} for the convolution algorithm has been found to be $-(30/40)$ dB.

A similar result is obtained using the convolution algorithm with a high interpolation order (8–16). The computation complexity of the first algorithm is about N^3 ($N \times N$ is the number of pixels on the image) and that of the second algorithm is about kN^2 (k is a constant varying in proportion with the interpolation order). The computation time with the convolution algorithm is 3–5 times longer than with the interpolation algorithm. Its application is, however, preferred because it allows processing primary data as they come handy (e.g. the internal integral in Eq. (3.74)) in real time for each projection individually. The convolution algorithm can be used for simultaneous (systolic) computations by a set of elementary processors such as a multiplier, a summator and a saving register, which are not tightly coupled to one another.

There have been some attempts to design 'faster' tomographic algorithms, using, for example, the Hankel transform. The principle of this algorithm is as follows. Because the functions $g(\rho, \Phi)$ and $G(r, \vartheta)$ are periodic with the period 2π, they can be expanded into a Fourier series:

$$g(\rho, \Phi) = \sum_{n=-\infty}^{\infty} g_n(\rho) \exp(jn\Phi),$$

$$G(r, \vartheta) = \sum_{m=-\infty}^{\infty} G_m(r) \exp(jm\vartheta),$$

where

$$g_n(\rho) = \frac{1}{2\pi} \int_{-\pi/2}^{\pi/2} g(\rho, \Phi) \exp(-jn\Phi) d\Phi,$$

$$G_m(r) = \frac{1}{2\pi} \int_{-\pi/2}^{\pi/2} G(r, \vartheta) \exp(-jm\vartheta) d\vartheta.$$

In addition, We can show that

$$g_n(\rho) = 2\pi \int_0^{\infty} r G_n(r) J_n(r\rho) dr, \qquad (3.75)$$

where $J_n(\cdot)$ is the first-order Bessel function. This relation is known as the nth order Hankel transform [103].

Apparently, these relations can be applied to the reconstruction of g from the known values of G. An important advantage of this algorithm is the use of data in a polar format without interpolation. The Hankel transform takes the largest computational time. The available procedures for accelerating the computation are based on the representation of Eq. (3.75) as a convolution and the use an asymptotic representation of the Bessel function.

The available tomographic algorithms for image reconstruction in spotlight SAR also include signal processing designs accounting for the wave front curvature. These employ more complex transformations than just finding Fourier images. The 'efficiency' of such algorithms should be evaluated taking into account the inadequacy of the problem formulation. We should recall that a problem is considered to be ill-posed if it has no solution, or the solution is ambiguous or unsteady, that is, it does not change continuously with the input data. It is the second circumstance that usually takes place in the case being discussed, because experimental data fit only a small region in the transformation space. Even if we assume that the $G(X, Y)$ values are known over the whole polar grid, there is generally no sampling theorem for $g(\rho, \Phi)$ in the polar format.

The tomographic approach allows estimation of all major parameters of the spotlight SAR. In particular, the resolution was estimated as

$$\delta_x \cong \frac{\pi c}{2\alpha T},$$
$$\delta_y \cong \frac{\pi c}{2\omega_0 \sin(|\vartheta_{\min}| + |\vartheta_{\max}|)}$$

a value coinciding with a conventional radar estimate [100]. The conditions for the input data discretisation were defined. Besides, requirements for the synthesis were formulated, providing that one could ignore the deviation of projections from a straight line and their incoherence due to the wave front curvature in the target vicinity.

We have made the above analysis for a 2D case, neglecting the SAR's altitude. This circumstance does not, however, violate the generality of our treatment. A correction for the altitude can be easily made by 'extending' the linear range by a factor of R_o/R, where R_o is the slant range to the target's centre and R is the slant range projection onto the earth plane.

We should like to emphasise the following important difference between CAT systems and SAR operating in a spot-light (telescopic) mode. In order to provide a high resolution, a CAT radar must cover a much larger range of angles than a SAR, say, 360° against 6°. This can be understood in terms of image reconstruction from data obtained within a limited region of a 2D space–time spectrum. In this sense, the spectral region utilised by the SAR is shifted relative to the origin by $2\omega_0/c$ (Eq. (3.71)), while the spectral region of a CAT radar is not. We shall try to show why a high resolution can be achieved by a small aperture in SAR.

We should first recall that resolution corresponds to the width of the major lobe of the pulse response, normally at 3 dB. The resolving power of both CAT and SAR systems depends only on the frequency band used in a 2D spectrum and it should be independent of the carrier frequency ω_0, which is the frequency of the band shift.

To illustrate, the range resolution for the shaded region in Fig. 1.7 is inversely proportional to the frequency band width along the X-axis (or the u-axis) and the azimuthal resolution to that along the Y-axis (or the v-axis).

If the number of point objects is large, the image quality becomes poor due to signal interference. This effect arises because the pulse response of the system, usually expressed as a 2D function $\sin x/x$, contains a constant phase factor varying with the carrier frequency ω_0 and the position of the point object. As is easy to see, the quality of a reconstructed image is independent of the ω_0 variation provided that the function describing the object depends on a complex-valued variable with an occasional uncorrelated phase. This means that the phases of signals reflected by different scattering centres are not correlated. The authors evaluated the image quality from a formula similar to that for finding a mean-square root error. One can suggest that the process of SAR imaging meets this condition. As a result, the spectrum of the 'initial image' occupies a wide frequency band in Fourier transform space and the object's reflectivity can be reconstructed from a limited shifted spectral region. This circumstance is similar to a fact well known in holography: the image of a diffusely scattering object can be reconstructed from any fragment of the hologram (Chapter 2).

These aspects of image quality can be treated in a different way. The band width of space frequencies, Δv, which defines the azimuthal resolution, 'grows' with the shift frequency (Fig. 1.7) as

$$\Delta v = (|\vartheta_{\min}| + |\vartheta_{\max}|)(2\omega_0/c).$$

For a CAT radar, $\omega_0 = 0$ and Δv is

$$(|\vartheta_{\min}| + |\vartheta_{\max}|)\Delta u,$$

where $\Delta u \cong 4\alpha T/c \ll 2\omega_o/c$. Therefore, in order to obtain a high azimuthal resolution, one must have information about the whole range of view angles, 360°.

One can eventually say that the principal difference between the CAT and SAR systems is that the latter is coherent and can process complex signals.

To conclude, the tomographic principle of synthetic aperture operation does not rely on the analysis of Doppler frequencies of reflected signals. We shall turn to this factor again in Chapters 5 and 6 when we describe imaging of a rotating object by an inverse synthetic aperture. It will be shown that the holographic and tomographic approaches do not need an analysis of Doppler frequencies.

Chapter 4
Imaging radars and partially coherent targets

Remote sensing of the earth surface in the microwave frequency range is a rapidly developing field of fundamental and applied radio electronics [31,77]. It has already become a powerful method in many earth sciences such as geophysics, oceanology, meteorology, resources survey, etc. Especially among microwave sensors side-looking synthetic aperture radars (SAR) are capable of providing high-resolution images of a background area at any time, irrespective of weather conditions. Extensive information has been obtained by airborne radars and radars carried by satellites and spacecraft: SEASAT-A and SIR (USA), RADARSAT (Canada), Almaz-1 (Russia), ERS and ENVISAT (European Space Agency), Okean (Russia, Ukraine). A challenge to the radar scientist is the analysis of synthetic aperture imaging of extended targets.

The various tasks of remote SAR sensing of the earth include the study of the ocean surface, sea currents, shelf zones, ice fields, and many other problems [62]. Objects to be imaged are wind slicks, oil spills, internal waves, current boundaries, etc. Some of these targets are characterised by motions with unknown parameters, so they are considered to be partially coherent. This chapter focuses on theoretical problems of SAR imaging of such targets while their practical aspects are discussed in Chapter 9.

In contrast to a conventional radar which measures instantaneous amplitudes of a signal reflected by a target, the SAR registers the signal phase and amplitude for a finite synthesis time T_s. The conversion of these data to a radar image requires the knowledge of the time variation of these characteristics, which can be found if one knows *a priori* the time variation of the reflected signal. When the view zone includes only stationary targets, the prescribed data have the form of the time dependence of distances between the SAR and the objects being viewed. If the time variation of the signal phase is unknown, the coherence is violated. This may happen not only in SAR viewing of the sea surface but also because of sporadic fluctuations of the carrier trajectory (see Chapter 7). So partial coherence may be associated with the viewing conditions or with the target itself. The analytical method discussed below preserves its generality in this case.

4.1 Imaging of extended targets

Viewing of background surfaces by SAR involves two kinds of difficulty: one is associated with evaluation and improvement of image quality and the other with image interpretation [59]. The first difficulty is due to the fact that one has to control the SAR performance (i.e. the operation of transmitters/receivers and imaging devices), to evaluate the capabilities of test systems, and to compare the data from the synthetic aperture and other sensors. The other difficulty arises from the diversity of image applications. The point is that one resolution element contains a large number of elementary scatterers reflecting coherent signals which interfere with one another. This produces speckle noise on the radar image. The situation becomes especially complicated, for example, in sea surface viewing when elementary scatterers move, making the image intensity a random quantity. For this reason, one has to employ statistical methods to describe the imaging of extended proper targets. It is clear that both problems are closely interrelated. For instance, the statistical characteristics of speckle noise can be used to obtain information about the surface and to evaluate the image quality.

Image quality is affected by numerous independent parameters of target imaging. Therefore, image evaluation requires the use of quantitative factors which can objectively describe the image characteristics and relate this information to the parameters of the viewing system. The quality of any image, including a radar one, can be described by four parameters: geometrical accuracy, spatial resolution, radiometric precision and radiometric resolution.

Geometric accuracy defines the longitudinal and latitudinal precision of the image as an integral entity, which is particularly important for images of poorly recognisable surface areas. It also determines the mapping accuracy of different points on the image relative to one another.

Since a SAR is a coherent system, its ability to resolve neighbouring point scatterers depends on various factors, such as the relative phases of the scatterers, their relative effective cross sections, the system noise, etc. So it is reasonable to describe spatial resolution either with the half width of the major impulse response peak (usually, at 3 dB) or with the envelope of this response. The latter way enables one to find the extent to which the image is affected by the side lobes of the impulse response, which are comparable with the major peaks of responses from neighbouring, less intensive scatterers that can be erroneously taken for images of independent point targets. Spatial resolution can be evaluated by a photometric study of the image of a bright point object, say, of a corner reflector, or by determining the amplitude image profile of an object with a sharp reflectivity variation, followed by the calculation of the impulse response from this profile gradient. The second approach is more accurate because the resolution evaluation is less affected by the limited dynamic range of the aperture.

Radiometric precision indicates to what extent the various brightness levels of the image reproduce the reflectivity variation of the radar target at particular wavelengths, polarisations and radiation incidences. To measure the radiometric precision, one can use calibrated extended targets with different values of the specific cross-section (SCS).

Radiometric (contrast) resolution characterises the ability to discern the SCS values of neighbouring elements and is largely determined by random signal fluctuations registered on the image. Such fluctuations may arise along the signal pathway from aperture or speckle noise. The radiometric resolution for homogeneous areas can be calculated from the density distribution function of the image intensity probability.

The reflectivity distribution across the area of interest is often assumed to be normal. Then the amplitude distribution of the reflected signal is described by Reighley's formula while the phase is taken to be uniform in the range from 0 to 2π. The radar image intensity, which is equal to the squared signal modulus, has an exponential distribution:

$$p(\chi) = (1+S)^{-1} \exp[-\chi/(1+S)], \tag{4.1}$$

where χ is the intensity normalised to unit noise power and S is the signal-to-noise ratio on the image. The average intensity and the distribution dispersion are, respectively, described as

$$\chi_m = 1 + S, \tag{4.2}$$

$$D_\chi = (1+S)^2. \tag{4.3}$$

SCS measurements involve a large ambiguity. From Eqs (4.2) and (4.3) it follows that the standard deviation of the SCS value is equal to the image intensity. To estimate this value, it is necessary to find the mean noise intensity and subtract it from the image intensity. We assume $\chi_m = S$ and assume that the estimate dispersion to be constant. If the radiometric resolution γ is found to be on the level of one standard deviation (the ratio of the mean value plus one standard deviation to the mean value), then for the distribution described by Eq. (4.1) at zero noise we have

$$\gamma = 10 \lg(2 + 1/S). \tag{4.4}$$

Obviously, γ will not be larger than 3 dB even at $S \to \infty$. The simplest way to improve radiometric resolution is to average the viewing results on several neighbouring resolution elements of an extended target (incoherent signal integration). Then we shall have

$$\gamma = 10 \lg[1 + (1+S)/N^{1/2}S], \tag{4.5}$$

where N is the number of uncorrelated integrated versions of the image.

Incoherent signal integration by SAR can be provided only at the expense of spatial resolution because this is normally done by multi-ray processing or by averaging the intensities of elements of a highly resolved image. For example, the SEASAT-A aperture used a four-ray processing which, nevertheless, could not totally remove the speckle noise [99].

Thus, there is a certain contradiction between spatial and radiometric resolutions [61]. A possible compromise is to choose a proper criterion for image quality. However, this is not very easy to do for two reasons. First, such a criterion must account for specific features of the object being viewed, which may happen to be diverse. Second, one must adapt this criterion to the subsequent processing of the image – visual, automated, etc. Moore [99], for example, suggested using visual expertise of the image as a criterion for evaluation of its quality. For a quantitative analysis he used the spatial grey-level (SGL) volume $V = V_a V_R V_g(N)$, where V_a and V_R are the azimuth and range resolutions, respectively, and $V_g(N)$ is the grey-level resolution defined by the number of uncorrelated integrated realisations, N.

Before proceeding with the discussion of criteria that can optimise the coherent-to-incoherent signal ratio in the synthetic aperture, we think it is necessary to consider briefly the available methods to describe SAR mapping of a typical fluctuating extended target – a rough sea surface.

4.2 Mapping of rough sea surface

At present we have much information on rough sea surface viewing by SAR systems [36,62], both airborne and carried by spacecraft. Most of the publications describe wave movements and their effect on radar image quality. However, this issue still remains controversial and is a subject of much debate [56].

When the sea surface is viewed by an airborne or space SAR, the probing radiation incidence varies from 20° to 70°. Bragg scattering by small-scale and capillary waves has the greatest effect on the reflection of electromagnetic radiation. The effect of large-scale (gravitational) waves on the radar image reveals itself in the modulation of scattering by small-scale waves. These phenomena are usually described by a 2D model which considers the sea surface as a superposition of Bragg scatterers – capillary and longer gravitational waves. They can also be described by a facet model, in which facets represent small-scale scatterers with superimposed capillary waves; the scatterers move with orbital velocities defined by large-scale waves [59]. The imaging of large-scale waves is affected by the following factors:

- the energy modulation of capillary waves due to hydrodynamic interaction between capillary and gravitational waves;
- the modulation of the facet inclination, which changes the effective incidence of the probing signal with respect to the normal facet surface, which, in turn, changes the Bragg scattering coefficient;
- the variations in the facet parameters (the position and the normal direction) and the Bragg scattering coefficient due to the facet movement during the synthesis.

The first two processes are important for sea viewing by any radar, whereas the third process affects only SAR imaging. The effect of moving waves on the image quality can be found analytically if one bears in mind that the synthesis time (0.1–3 s) is much shorter than the period of a large-scale wave (8–16 s). Then the functions that describe the time variation of the facet parameters and scattering coefficients can be expanded into a Taylor series. The major expansion terms are related to the radial components (along the slant range) of the orbital velocity and acceleration of the facets. These components are responsible for two effects: the velocity bunching and the image defocusing along the azimuth. The velocity bunching is associated with the azimuthal shift of each facet image because of the radial velocity effect, which represents a periodic rarefaction and thickening of virtual positions of elementary scatterers along the large-scale wave pattern. The bunching degree varies with the number of images of individual facets per unit azimuthal length, which is proportional to

$$\Delta = \frac{R}{v}\frac{du_r}{dx}, \qquad (4.6)$$

where R is the inclined range, v is the SAR carrier velocity, u_r is the radial velocity component and x is the azimuthal coordinate on the sea surface. For small values of $|\Delta|$, this effect is linear and is characterised by a linear transfer function; for large $|\Delta|$ values ($>\pi/2$), it becomes nonlinear, leading to image distortions. It is greatest for waves running along the azimuthal coordinate but practically vanishes for radial waves.

Image defocusing of large-scale waves is interpreted as being either due to the radial acceleration of the facets or due to the change in the relative aperture velocity because of the effect of the azimuthal phase velocity of sea waves [61]. Investigations have shown that the latter explanation is better substantiated. The major contribution to the image is made by the amplitude modulation of the reflected signal due to the surface roughness and facet inclination, whereas the velocity bunching plays a minor role. As for the image defocusing, it can be removed by correcting the signal processing conditions, for example, by an additional adjustment of the optical processor or by refining the base function during digital image reconstruction.

Generally, the sea wave behaviour appears to be quite complex. For this reason, available models of a probing signal reflected by the sea surface depend on the particular problem to be solved. Models accounting for the orbital motion of liquid droplets are too sophisticated to be extended to a large class of objects defined as partially coherent. Besides, they do not readily apply to the analysis of the influence of aperture parameters on image quality, because imaging is then determined only by the sea wave characteristics and viewing geometry. Probably, the only factor that affects the sea imaging by SAR and related to the choice of radar parameters is the image defocusing. But even here, we deal with the mapping of sea waves, which is a particular problem that does not represent the whole class of partially coherent targets.

On the other hand, of academic interest and practical importance are the problems of background dynamics, various anomalies in the extended target reflectivity (for the sea, these are slicks, spills of surface-active substances, etc.), as well as the proper choice of the SAR design for viewing this class of targets. The analysis shows that the results obtained can be extended to a large number of partially coherent extended targets.

In principle, the basic characteristics of extended target images, including images of sea surface, could be found by solving the problem of electromagnetic wave scattering by a moving plane. The methods of dealing with these problems are well known but they involve cumbersome calculations.

Another way of describing a radar signal reflected by an extended target is to introduce the autocorrelation function for the object being viewed, as is done in optical systems theory [29]. In this approach, a complex signal reflected by the sea surface can be written as $U(x,t) = u(x,t)u_r(x,t)$, where $u(x,t)$ is a co-factor accounting for the effect of large-scale sea waves and $u_r(x,t)$ is a random complex component to describe the signal reflected by a capillary wave. The autocorrelation function of this signal is

$$\langle U(x_1,t_1)U^*(x_2,t_2)\rangle = u(x_1,t_1)u^*(x_2,t_2)\langle u_r(x_1,t_1)u_r^*(x_2,t_2)\rangle,$$

where the asterisk denotes the complex conjugate and $\langle\rangle$ represents the ensemble average.

The complex component $u_r(x,t)$ can be written as

$$u_r(x,t) = f(x)\alpha(t\mid x) \tag{4.7}$$

where $f(x)$ is a complex random amplitude of the reflected signal, defined by the surface roughness, and $\alpha(t\mid x)$ is a complex reflectivity describing the time fluctuations of the reflected signal with the x-coordinate.

Normally, $f(x)$ describes the Gaussian random process with a zero average, which happens in the case of Bragg scattering of an electromagnetic wave on a rough surface. The spatial correlation function of this process can be approximated by the Dirac delta-function when the spacing between the features is sufficiently small, a condition often fulfilled in practice:

$$\langle f(x_1)f^*(x_2)\rangle = p\delta(x_1 - x_2), \tag{4.8}$$

where p is a factor proportional to the object's SCS and is defined by the governing radar equation.

The autocorrelation function of the time fluctuations of the surface is, in turn, equal to

$$\langle \alpha(t_1\mid x_1)\alpha^*(t_2\mid x_2)\rangle = \Gamma[(t_1 - t_2)\mid x_1, x_2]. \tag{4.9}$$

It has been termed partial or autocorrelation coherence [103]. The possibility to employ this formalism is a fundamental feature of partially coherent objects which can then be treated as a special class of targets.

Thus, the autocorrelation function of the signal reflected by the sea surface can be written as

$$\langle U(x_1,t_1)U^*(x_2,t_2)\rangle = u(x_1,t_1)u^*(x_2,t_2)\delta(x_1 - x_2)\Gamma[(t_1 - t_2)\mid x_1, x_2]. \tag{4.10}$$

Taking the time fluctuations of the signal to be constant, we can approximate the autocorrelation function with the expression

$$\Gamma[(t_1 - t_2)\mid x_1, x_2] = \exp[-\pi(t_1 - t_2)^2/\tau_c^2], \tag{4.11}$$

where τ_c is the time interval of the correlation.

The radar signal model discussed above agrees well with experimental data [112]. Equation (4.11) has a general form allowing the solution of a large range of problems involved in the analysis of extended target imaging by SAR systems. We shall further omit partially coherent background modulation by large-scale waves, assuming $u(x,t) = 1$ in order to be able to extend the results to a sufficiently large class of objects.

The model we have described can provide the basic statistical characteristics of partially coherent surface images, but we should first outline the imaging model itself.

4.3 A mathematical model of imaging of partially coherent extended targets

Suppose a SAR is borne by a carrier moving uniformly along a straight line with a velocity v. The carrier position is described by the coordinate $y = vt$ and inclined range R, while the position of an arbitrary element of the viewed surface is described by the x-coordinate (Fig. 4.1). The imaging process is subdivided into two stages – the registration of the reflected signal (hologram recording) and the image reconstruction. This approach allows one to represent a general block diagram of the synthetic aperture (Fig. 4.2) with the complex amplitude of the reconstructed image written as a sum of convolutions:

$$s = f * w * h + n * h, \tag{4.12}$$

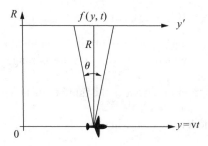

Figure 4.1   The geometrical relations in a SAR

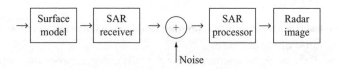

Figure 4.2   A generalised block diagram of a SAR

where f is a function of the viewed surface reflectivity; w and h are the impulse responses of the radar and the processor, respectively; n is the complex amplitude of additive noise; and $*$ denotes convolution.

The optimal quality of images of point objects is achieved by matching the impulse responses of the radar and the aperture processor:

$$h(y) = w^*(y). \tag{4.13}$$

This condition cannot, however, provide an optimal image of an extended proper object [99], since it is impossible to integrate an incoherent signal and to reduce the speckle noise on the image. On the other hand, the fact that the image intensity $g(u) = s(u)s^*(u)$ is usually registered at the aperture processor output allows introducing the concept of a partially coherent processor in square filtration theory [58]. One can then account simultaneously for the effects of coherent and incoherent signal integration by the aperture and eventually obtain the major statistical characteristics of images of partially coherent extended targets. This type of processor will have the following impulse response:

$$Q(y_1, y_2) = \gamma(y_1 - y_2)h(y_1)h^*(y_2), \tag{4.14}$$

where $\gamma(y_1 - y_2)$ is a factor characterising the degree of incoherent signal integration. Then Eq. (4.13) will be valid for any class of targets.

To avoid cumbersome calculations, we shall introduce Gaussian approximations of the functions

$$w(y) = \exp(-ay^2/2)\exp(jby^2/2), \tag{4.15}$$

$$h(y) = \exp(-ay^2/2)\exp(jby^2/2), \tag{4.16}$$

$$\gamma(y_1 - y_2) = \exp[-A(y_1 - y_2)^2/2], \tag{4.17}$$

where $\int \exp(-ay^2/2)dy = (2\pi/a)^{1/2} = \Theta R$ is the width of the real antenna pattern projection onto the area, which defines the synthesis range $b = 2\pi/\lambda R$; λ is the aperture wavelength; and $\exp(-Lx^2/2)\,dx = (2\pi/L)^{1/2} = L_s$ is the synthesised aperture length. The A/a ratio describes the number of independent integrations of an incoherent signal, $N = (1 + A/a)^{1/2}$.

Within this SAR model, the image intensity is

$$g(u) = \int\!\!\!\int_{-\infty}^{\infty} Q(u - y_1, u - y_2)[s_r(y_1) + n(y_1)]\,[s_r^*(y_2) + n^*(y_2)]\,dy_1\,dy_2, \tag{4.18}$$

where $s_r(y)$ describes the complex hologram function and $n(y)$ is a function describing the intrinsic noise of the aperture.

The model of a synthetic aperture with a partially coherent processor can be used to analyse statistical characteristics of images of partially coherent targets and to reveal the effects of coherent and incoherent signal integration on the image parameters.

4.4 Statistical characteristics of partially coherent target images

Let us turn back to the synthetic aperture shown in Fig. 4.1. In one of the range channels, the reflected signal can be represented as a random complex field. For many real surfaces, the function $f(y)$ in the centimetre wavelength range is a Gaussian random process with the zero average and the correlation function in the form of the Dirac delta-function obeying Eq. (4.8).

The time relations for the surface changes can be described by the autocorrelation function of Eq. (4.9) and that of the reflected signal, assuming $u(y_1, t_1) \equiv s_0(y_1, t_1)$:

$$\langle s_0(y_1, t_1) s^*_0(y_2, t_2)\rangle = p\delta(y_1 - y_2)\Gamma[(t_1 - t_2) \mid y_1, y_2], \tag{4.19}$$

where the function $\Gamma[(t_1 - t_2) \mid y_1, y_2]$ is defined by Eq. (4.11).

The process of imaging can be analysed in terms of the holographic approach, as applied to SAR. At the first stage, the hologram is recorded: $s_h = s_0 * w + n$, where w is the impulse response of the aperture receiver, n is additive noise, and $*$ denotes convolution. At the second stage, the image is reconstructed with the intensity

$$g = ss^* = (s_h * h) * (s_h * h)^*, \tag{4.20}$$

where s is the complex amplitude of the image and h is the impulse response of the aperture processor.

To smooth out the image fluctuations, one usually uses incoherently integrated signals. We can now evaluate the effects of two smoothing procedures: multi-ray processing and averaging of neighbouring resolution elements on the image [2]. Additionally, we shall consider the potentiality of incoherent signal integration on the hologram. In the first case, when the image is reconstructed by an optical processor, its intensity is [60]

$$g_1(u) = \int g(u, \tau) D_a(\tau) \, d\tau. \tag{4.21}$$

Here $D_a(\tau)$ describes the light distribution across the aperture stop located in front of the secondary film which records the image, τ is the current exposure of the secondary film, and u is the reconstructed image coordinate.

In the second case, the image intensity is

$$g_2(u) = \int g(u') G_a(u - u') \, du', \tag{4.22}$$

where $G_a(u - u')$ is the weighting function of the averaging.

In the third case, the image intensity is given by Eq. (4.20), in which the hologram function is

$$s_{ha}(y') = \int s_h(y'') C_a(y' - y'') \, dy'', \tag{4.23}$$

where $C_a(y' - y'')$ is the weighting function of the averaging and $y' = vt$ is the spatial coordinate on the hologram.

88   Radar imaging and holography

To simplify the calculations, let us approximate the above functions with the expressions

$$D_a(\tau) = \exp[-D(\tau v)^2/2],$$
$$G_a(u) = \exp[-Gu^2/2], \quad (4.24)$$
$$C_a(y) = \exp[-Cy^2/2],$$

where $\int \exp(-Dv^2\tau^2/2)d(v\tau) = D_e$ is the equivalent width of the aperture stop, $(2\pi/G)^{1/2} = G_e$ and $(2\pi/C)^{1/2} = C_e$ are the equivalent widths of the respective weighting functions; Eqs (4.15) and (4.16) have been used as approximations of the functions $w(y)$ and $h(y)$.

We can now find the following parameters characterising the statistical properties of the image: the average intensity g_a, the intensity dispersion σ_a^2, the smoothing degree g_a^2/σ_a^2, the autocorrelation range u_c and the signal-to-noise ratio $W_g = g_a/g_{an}$, where g_{an} is the average noise on the image.

4.4.1 Statistical image characteristics for zero incoherent signal integration

The parameters of interest can be found using the power spectrum of the image intensity [60]:

$$S_g(\omega) = (2\pi)^{-1} \int |H(\eta, \omega - \eta)|^2 S_h(\omega) S_h(\omega - \eta) \, d\eta, \quad (4.25)$$

where $H(\eta, \omega)$ is a 2D transfer function of the aperture processor and $S_h(\omega)$ is the hologram power spectrum. In turn, $H(\eta, \omega) = H(\eta)H^*(\omega)$, where $H = F\{h\}$ is the Fourier image of the function $h(x)$. With Eq. (4.24), we get

$$H(\eta, \omega) = (L^2 + b^2)^{-1/2} \exp\left[-(\eta^2 + \omega^2)\frac{L}{2(L^2 + b^2)}\right]$$
$$\times \exp\left[j(\omega^2 - \eta^2)\frac{b}{2(L^2 + b^2)}\right]. \quad (4.26)$$

The function $S_h(\omega)$ represents the Fourier transform of the hologram spatial correlation function, $R_h(\Delta y')$, which can be described, for low intrinsic aperture noise, as

$$R_h(\Delta y') = s_h(y_1')s_h^*(y_2') = p(\pi/a)^{1/2} \exp[-(y_1' - y_2')^2(a^2 + b^2 + 2aB)/4a] \quad (4.27)$$

with

$$\Delta y' = y_1' - y_2', \quad B = 2\pi/(v\tau_c)^2.$$

Hence, we have

$$S_h(\omega) = p[2\pi/(a^2 + b^2 + 2aB)]^{1/2} \exp[-\omega^2/(a^2 + b^2 + 2aB)]. \tag{4.28}$$

By substituting Eqs (4.26) and (4.28) into Eq. (4.25) and using the expression $\text{cov}_g(u) = F\{S_g(\omega)\}$ for the background, we obtain

$$<g_a> = \int S_g(\omega) d\omega = 2^{1/2}\pi p[aL(a+L+2B) + b^2(a+L)]^{-1/2},$$

$$\sigma_g^2 = \text{cov}_g(0) = 2(\pi p)^2 [aL(a+L+2B) + b^2(a+L)]^{-1},$$

$$\langle u_c \rangle = \int \text{cov}_g(u)/\text{cov}_g(0) du$$

$$= \pi[2a/(a^2 + b^2 + 2aB) + 2L/(L^2 + b^2)]^{1/2},$$

$$<g_a^2>/\sigma_g^2 = 1. \tag{4.29}$$

Assuming that the spectrum of the intrinsic aperture noise recorded on the hologram is uniform and has spectral density $S_{hn}(\omega) = n$, we find the respective parameters of the image noise:

$$<g_{an}> = n(\pi/L)^{1/2},$$

$$\sigma_n^2 = n^2(\pi/L),$$

$$\langle u_{cn} \rangle = \pi[2L/(L^2 + b^2)]^{1/2},$$

$$g_{an}^2/\sigma_n^2 = 1. \tag{4.30}$$

The signal-to-noise ratio $W_h = g_a/g_{an}$ can be reduced to $W_h = W_0 Q$, where W_0 is a classical quantity and

$$Q = [(a/b^2 + 1/L)(a+L) + 2aB/b^2]^{-1/2} \tag{4.31}$$

is a factor largely determined by the real antenna pattern.

A quantitative analysis of Eqs (4.29) and (4.30) shows that the statistical parameters of the image are practically independent of the surface fluctuations at the typical values of $\lambda \approx 3$ cm, $R \approx 10\text{--}20$ km, $\Theta \approx 0.02$ and $\tau_c \approx 0.01$ s and that the correlation ranges $\langle u_c \rangle$ and $\langle u_{cn} \rangle$ differ only slightly with a maximum at $L_s = (\lambda R/2)^{1/2}(b = L)$. The latter circumstance can be attributed to the fact that the function $h(y)$ essentially represents a linearly frequency-modulated signal, whose spectral width is proportional to its range at $L_s > (\lambda R/2)^{1/2}$ and inversely proportional at $L_s < (\lambda R/2)^{1/2}$. So the spectral width of the image fluctuations is minimal at $L_s = (\lambda R/2)^{1/2}$.

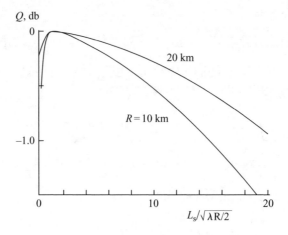

Figure 4.3 *The variation of the parameter Q with the synthesis range L_s at $\lambda = 3$ cm, $\Theta = 0.02$ and various values of R*

At the minimal width of the $|H(\omega)|$ function, the difference between $\langle g_a \rangle$ and $\langle g_{an} \rangle$ is also insignificant. This accounts for the maximum of the Q function at $L_s = (\lambda R/2)^{1/2}$ (Fig. 4.3). A quantitative analysis of Q shows that the influence of the real aperture pattern on the signal-to-noise ratio is slight and reveals itself only at large synthesis ranges, $L_s \ll (\lambda R/2)^{1/2}$.

4.4.2 Statistical image characteristics for incoherent signal integration

According to Eq. (4.21), the image intensity in multi-ray processing is [60]

$$g(u) = \int D_a(v\tau) \iint s_h(y_1') s_h(y_2') \exp[-L(y_1' - u - v\tau)^2/2]$$
$$\times \exp[-L(y_2' - u - v\tau)^2/2] \times \exp[jb(y_1' - u)^2/2]$$
$$\times \exp[jb(y_2' - u)^2/2] \, dy_1' \, dy_2' \, d(v\tau). \tag{4.32}$$

This relation describes the impulse response of the aperture processor and enables one to find its transfer function:

$$H(\eta, \omega) = r(l^2 + b^2 + 2Al)^{-1/2}$$
$$\times \exp\{-[A(\eta + \omega)^2 + l(\eta^2 + \omega^2)]/[2(l^2 + b^2 + 2Al)]\}$$
$$\times \exp\{-jb(\eta^2 - \omega^2)/[2(l^2 + b^2 + 2Al)]\}, \tag{4.33}$$

with

$$r = 2\pi/(D + 2L)^{1/2}, \quad l = LD/(D + 2L), \quad \text{and} \quad A = L^2/(D + 2L).$$

Following the same procedure and using the last two relations, we can find the characteristics of the background and noise on the image:

$$<g_a> = 2^{1/2}\pi p\{D[aL(a+L+2B)+b^2(a+L)]+2aLb^2\}^{-1/2}, \quad (4.34)$$

$$\sigma_g^2 = 2(\pi p)^2\{[L(D+L)(a^2+b^2+2aB)+aD(L^2+b^2)+2aLb^2]^2 \\ - L^4(a^2+b^2+2aB)\}^{-1} \quad (4.35)$$

$$\langle u_c \rangle = 2^{1/2}\pi\{L(1+2L/D)/[L^2+b^2(1+2L/D)] \\ + a/(a^2+b^2+2aB)\}^{1/2}, \quad (4.36)$$

$$<g_a^2>/\sigma_g^2 = \{1+2L^2b^2/[aD(L^2+b^2)+LDb^2+2aLb^2]\}^{1/2}, \quad (4.37)$$

$$<g_{an}> = \pi n[2/(LD)]^{1/2}, \quad (4.38)$$

$$\sigma_n^2 = \pi^2 n^2(1+2L/D)^{-1/2}/(LD), \quad (4.39)$$

$$\langle u_{cn} \rangle = 2^{1/2}\pi\{L(1+2L/D)/[L^2+b^2(1+2L/D)]\}^{1/2}, \quad (4.40)$$

$$<g_{an}^2>/\sigma_n^2 = (1+2L/D)^{1/2}. \quad (4.41)$$

The analysis of these relations shows that the image smoothing is improved, as was expected, while the correlation functions of the clutter and radar noise images are practically the same, $\langle u_c \rangle \approx \langle u_{cn} \rangle$. Figure 4.4 demonstrates the correlation range versus the normalised quantity L_s for various degrees of incoherent integration D_e, or for

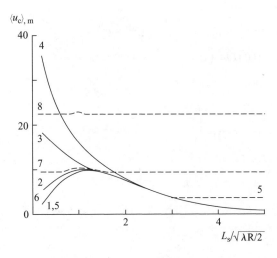

Figure 4.4   The dependence of the spatial correlation range of the image on normalised L_s for multi-ray processing (solid lines) at various degrees of incoherent integration D_e and for averaging of the resolution elements (dashed lines) at various G_e; $\lambda = 3\,cm$, $R = 10\,km$; 1, 5–0 (curves overlap); 2, 6–0.25$(\lambda R/2)^{1/2}$; 3, 7–$(\lambda R/2)^{1/2}$; 4, 8–2.25$(\lambda R/2)^{1/2}$

different aperture stop sizes. It is clear that the image correlation at $L_s > (\lambda R/2)^{1/2}$ (the focused processing region) will only slightly vary with D_e but the correlation range in incoherent integration will become larger (the defocused processing region). The parameter Q then takes the form

$$Q = [(a+L)(a/b^2 + 1/L) + 2aB/b^2 + 2a/D]^{-1/2}.$$

Its quantitative analysis indicates that it does not much affect the signal-to-noise ratio.

When the resolutions of neighbouring elements are averaged according to Eq. (4.22), the processor transfer function is expressed as

$$H(\eta, \omega) = r_1(l_1^2 + b_1^2 + 2A_1 l_1)^{-1/2}$$
$$\times \exp\{-[A_1(\eta+\omega)^2 + l_1(\eta^2 + \omega^2)]/[2(l_1^2 + b_1^2 + 2A_1 l_1)]\}$$
$$\times \exp\{-jb_1(\eta^2 - \omega^2)/[2(l_1^2 + b_1^2 + 2A_1 l_1)]\} \quad (4.42)$$

with

$$r_1 = [2\pi/(G+2L)]^{1/2}, \quad l_1 = LG/(G+2L),$$
$$A_1 = (L^2 + b^2)/(G+2L) \quad \text{and} \quad b_1 = bG/(G+2L). \quad (4.43)$$

Hence, we have

$$\langle g_a \rangle = 2^{1/2}\pi p\{G[a(L^2 + b^2) + L(a^2 + b^2 + 2aB)]\}^{-1/2}, \quad (4.44)$$
$$\sigma_g^2 = 2(\pi p)^2 \{G^2[Lb^2 + a(L^2 + b^2)]^2 + 2Gb^2(L^2 + b^2)$$
$$[Lb^2 + a(L^2 + b^2)]\}^{-1/2}, \quad (4.45)$$
$$g_a^2/\sigma_g^2 = \{1 + 2b^2(L^2 + b^2)/[aG(L^2 + b^2) + LGb^2]\}^{1/2}, \quad (4.46)$$
$$\langle g_{an} \rangle = \pi n[2/(LG)]^{1/2}, \quad (4.47)$$
$$\sigma_n^2 = \pi^2 n^2 \{LG[LG + 2(L^2 + b^2)]\}^{-1/2}, \quad (4.48)$$
$$\langle g_{an}^2 \rangle / \sigma_n^2 = [1 + 2(L^2 + b^2)/(LG)]^{1/2}, \quad (4.49)$$
$$\langle u_c \rangle = 2^{1/2}\pi\{[Lb^2 + a(L^2 + b^2)]/[b^2(L^2 + b^2)] + 2/G\}^{1/2}, \quad (4.50)$$
$$\langle u_{cn} \rangle = 2^{1/2}\pi[L/(L^2 + b^2) + 2/G]^{1/2}. \quad (4.51)$$

In this case, we also have $\langle u_c \rangle \approx \langle u_{cn} \rangle$. Figure 4.4 illustrates this dependence at various widths of the integrating function G_e. Obviously, the image correlation range increases in proportion with the integrating window width. The expression for the coefficient Q coincides with Eq. (4.31), since the statistical properties of the background and noise images are similar and cannot contribute to the power.

In incoherent signal integration on the hologram, the $H(\eta, \omega)$ function is described by Eq. (4.26) and the R_h function, after averaging by Eq. (4.23), takes the form:

$$R_h(\Delta y') = p[\pi/C(aC + a^2 + b^2 + 2aB)]^{1/2}$$
$$\times \exp[-(\Delta y')^2 C(a^2 + b^2 + 2aB)/4(aC + a^2 + b^2 + 2aB)].$$

Therefore, the noise correlation function on the hologram can be written as

$$R_{hn}(\Delta y') = n(\pi/C)^{1/2} \exp[-(\Delta y')^2 C/4. \tag{4.52}$$

This expression yields the hologram signal-to-noise ratio $W_h = R_h(0)/R_{hn}(0) = W_{h0} Q_h$, where W_{h0} is the single pulse ratio defined by the governing radar equation as $Q_h = N_i(1 + N_i^2/K_a)^{1/2}$, $K_a = d_a/(2vT_{ir})$, d_a is the horizontal dimension of the real antenna, T_{ir} is the repetition rate of the pulses, N_i is the number of pulses integrated by the hologram with an incoherent averaging. The variation of Q_h with N_i is shown in Fig. 4.5. One can see that incoherent integration is profitable only at $N_i \leq K_a$, which agrees well with the condition $C_e \leq \langle y'_{hc} \rangle = 2(\pi a)^{1/2}/b$, where $\langle y'_{hc} \rangle$ is the correlation range of the hologram. If the latter condition is fulfilled, the basic statistical characteristics of the image can be described by expressions similar to Eqs (4.29)–(4.31), which means that there is no image smoothing.

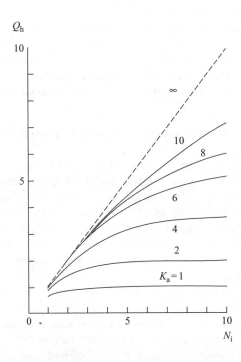

Figure 4.5 The variation of the parameter Q_h with the number of integrated signals N_i at various values of K_a

The results obtained allow the following conclusions to be made:

1. For typical conditions of SAR viewing of background surfaces and for real times $\tau_c \approx 0.01$ s of reflected signal correlation, all the image parameters discussed above are actually independent of the degree of coherence of the objects being viewed, in contrast to the radar resolving power.
2. The statistical properties of images of background surfaces and aperture noises are practically identical. This fact can be successfully used to calibrate radar apertures designed for measurement of background SCS. The maximum period of spatial image fluctuations is observed in the synthesis range $L_s = (\lambda R/2)^{1/2}$.
3. The analytical expressions we have derived can be used for the calculation of the image smoothing degree in the case of incoherent signal integration.
4. The signal-to-noise ratio of the image is nearly independent of the synthesised aperture length or on the incoherent integration range.
5. Incoherent integration on a hologram does not change the statistical characteristics of the image, that is, it does not lead to image smoothing, provided that the integrating function width is smaller than the hologram correlation range. Otherwise, there is no noticeable improvement of the signal-to-noise ratio on the hologram; therefore, the signal integration procedure becomes meaningless.
6. The methods of incoherent signal integration we have discussed (multi-ray processing and averaging of resolution elements) give similar results on the smoothing of image fluctuations. Multi-ray processing is performed automatically if the image is reconstructed by exposing the secondary film in an optical processor. In the case of digital reconstruction usually based on fast Fourier algorithms, the averaging of resolution elements is preferable because the algorithm performance is very effective when one has to process vast data. Application of special-purpose digital processors may improve the situation.

4.5 Viewing of low contrast partially coherent targets

The major SAR characteristics for viewing low contrast targets such as sea currents, wind slicks, oil spills, etc., are the spatial resolution and radiometric (contrast) resolution determined by the number of incoherent signal integrations [58]. It is clear that a proper choice of the proportion between spatial and radiometric resolutions (coherent and incoherent integration) will depend not only on the radar parameters but on the properties of the target to be viewed. So it is reasonable to consider optimisation of SAR performance in the context of partial coherence of signals reflected by an extended target.

Recall that the process of imaging includes two stages. First, the received signal is recorded on a radar hologram as $u(y') = w(y' - y)f(y)$, where $w(y' - y)$ is the impulse response of the aperture receiver, $f(y)$ is a function describing the spatial distribution of the target reflectivity, y is the coordinate in the viewed surface plane, $y' = vt$ is the SAR carrier coordinate, and t is current viewing time. Second, the image field is recorded: $g(y'') = u(y')h(y' - y'')$, where $h(y' - y'')$ is the impulse response of the aperture processor and y'' is the image coordinate.

Imaging can be described in terms of linear filtration theory. The concepts of a quadratic filter and a frequency-contrast characteristic (FCC) well known from optics can be used to present the image intensity:

$$S_I(\omega) = S_o(\omega) K_R(\omega) \tag{4.53}$$

where $S_o(\omega)$ is the space frequency spectrum of the SCS of the object and $K_R(\omega)$ is the FCC of the aperture.

For instance, if the average SCS of the background is σ_0, the distribution of a low contrast target is described by the function

$$\sigma(y) = \sigma_0[1 - m\exp(-y^2 A/2)], \tag{4.54}$$

where $m < 1$ is a factor defining the target's initial contrast $K_{in} = (1-m)/(1+m)$ with respect to the background, $A = 2\pi/l^2$ is a parameter related to the target's size l, and the aperture FCC is given by the expression

$$K_R(\omega) = \exp[-\omega^2/(2z)], \tag{4.55}$$

where z denotes its width. Then using Eq. (4.53), we can write the spatial distribution of the image intensity:

$$g(y'') = \sigma_0\{1 - m[z/(A+z)]^{1/2}\exp[-(y'')^2 Az/2(A+z)]\}. \tag{4.56}$$

Hence, the object's contrast on the image is

$$K_{out} = \{1 - m[z/(A+z)]^{1/2}\}/\{1 + m[z/(A+z)]^{1/2}\} \tag{4.57}$$

and its observable size is

$$l' = [2\pi(1/A + 1/z)]^{1/2}. \tag{4.58}$$

It is clear that the contrast and target size on the image become distorted but the knowledge of the explicit quantity $K_R(\omega)$ can give the real object's parameters.

For targets whose reflectivity varies with time randomly, the signal received by the aperture possesses a partial coherence and the hologram function $u(y')$ is no longer a convolution integral. In that case it would be unreasonable to use linear filtration theory. We shall show, however, that statistical methods and physical assumptions concerning the time fluctuations of objects' reflectivities can make this convenient formalism work successfully.

For this, we shall find the aperture response for a low contrast target ($m \ll 1$), whose reflectivity distribution is described by the function

$$f(y,t) = [1 + m\cos(\Omega y)]f(y)\alpha(t \mid y), \tag{4.59}$$

where Ω is a certain space frequency and $\alpha(t \mid y)$ is a random complex function describing the time fluctuations of the reflected signal.

The aperture FCC can be written as $K_R(\Omega) = K_{out}/K_{in}$, where $K_{in} = (\langle\sigma\rangle - \langle\sigma_{m=0}\rangle)/\langle\sigma_{m=0}\rangle \approx 2m$; $K_{out} = (\langle g\rangle - \langle g_{m=0}\rangle)/\langle g_{m=0}\rangle$; $\langle\sigma\rangle = f(0,0)f^*(0,0)$; $\langle\sigma\rangle$

and $\langle g \rangle$ are the average values of the target's SCS and image intensity, respectively. The correlation function of the field in Eq. (4.59) is defined as

$$R_f = \langle f(y_1,t_1)f^*(y_2,t_2)\rangle = [1 + m\cos(\Omega y_1)]\langle f(y_1)f^*(y_2)\alpha(t_1 \mid y_1)\alpha^*(t_2 \mid y_2)\rangle. \tag{4.60}$$

For many real surfaces, $f(y)$ in the centimetre wavelength range is a Gaussian process with a zero average and a correlation function in the form of the Dirac delta-function of Eq. (4.8). Assuming the time fluctuations of the signal to be a steady-state random process, we can use the approximation of Eq. (4.11). Together with Eq. (4.60) and $m \ll 1, y' = v t$, we shall have

$$R_f = [1 + m\cos(\Omega y_1) + m\cos(\Omega y_2)]\delta(y_1 - -y_2)\exp[-(y_1' - y_2')^2 B/2$$

with $B = 2\pi/(v\tau)^2$.

The average image intensity is

$$\langle g \rangle = \iint h(y_1')h^*(y_2')R_u(y_1',y_2')\,dy_1'\,dy_2', \tag{4.61}$$

where $R_u(y_1',y_2')$ is the correlation function of the hologram:

$$R_u(y_1',y_2') = \iint R_f w(y_1 - y_1')w^*(y_2 - y_2')\,dy_1'\,dy_2'.$$

Using the Gaussian approximations of the impulse responses in Eqs (4.15) and (4.16), we obtain, instead of Eq. (4.61), $\langle g \rangle = \langle g_0 \rangle + 2\langle g_\Omega \rangle$, where $\langle g_0 \rangle = (2)^{1/2}\pi[aL(a + L + (a+L)]^{-1/2}$ is the average intensity of the fluctuating background image and

$$\langle g_\Omega \rangle = m\langle g_0 \rangle$$
$$\times \exp[-(\Omega^2/4)((a+L)(a+L+2B)/aL(a+L+2B) + b^2(a+L))]. \tag{4.62}$$

For real viewing, we have $b^2 \gg aL$ and $b^2 \gg LB$, which reduces Eq. (4.62) to

$$K_{\text{out}} = 2m\exp[-\Omega^2(a+L+2B)/4b^2],$$
$$K_R(\Omega) = \exp[-\Omega^2(a+L+2B)/4b^2]. \tag{4.63}$$

There is a certain relationship between the FCC and the azimuthal resolution of the aperture. The latter can be found from the width of the averaged impulse response to a fluctuating point target:

$$\delta_a = \int \langle g(y'')\rangle/\langle g(0)\rangle dy''. \tag{4.64}$$

The signal reflected by this target can be prescribed as $f(y,y') = \delta(y)\alpha(y')$, where $\alpha(y')$ describes the time fluctuations of the signal, whose correlation properties are defined by Eq. (4.11). With Eqs (4.61) and (4.64), we get

$$\delta_a = [\pi(a+L+2B)/b^2]^{1/2}.$$

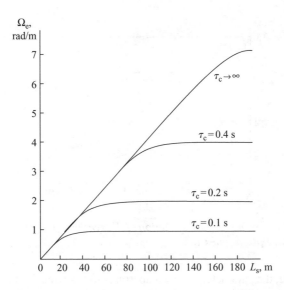

Figure 4.6 The variation of the parameter Ω_e with the synthesis range L_s at various signal correlation times τ_c

Of course, the aperture FCC can be presented as $K_R(\Omega) = \exp[-\Omega^2 \delta_a^2/(4\pi)]$ and its equivalent width as $\Omega_e = \int K_R(\Omega)\, d\Omega = 2\pi/\delta_a$.

The concept of FCC allows the consideration of a SAR as a linear filter of space frequencies. On the other hand, the filter description essentially depends on the target's behaviour through the parameter B. Figure 4.6 illustrates the variation of Ω_e with the synthesis range L_s for an airborne SAR. The basic radar parameters are $\lambda = 3$ cm, $R = 10$ km, $\Theta \approx 0.02$, and $v = 250$ m/s. For zero signal fluctuations ($\tau \to \infty$), the width Ω_e increases in proportion with L_s but at $L_s \approx \Theta R$ the linear dependence is violated because of the antenna pattern effect through the parameter a. The signal fluctuations lead to the resolution independence of L_s at $L_s > v\tau$ but they are rather defined by the correlation time τ. Equation (4.63) can be re-written in the form:

$$K_R(\Omega) = K_0(\Omega) K_\tau(\Omega), \tag{4.65}$$

where $K_0(\Omega) = \exp[-\Omega^2 (a+L)/(4b^2)]$ is the aperture FCC in the absence of signal fluctuations and $K_\tau(\Omega) = \exp[-\Omega^2 B/(2b^2)]$ is multiplicative noise arising from fluctuations in the radar channel.

Therefore, a SAR can be described as a set of two filters – a filter of space frequencies $K_0(\Omega)$ and a narrow band space–time filter $K_\tau(\Omega)$, whose bandwidth is determined by the time of the surface fluctuation correlation. The image has a spatial intensity spectrum $S_I(\Omega) = S_0(\Omega) K_R(\Omega)$. On the other hand, one can consider that the aperture measures the space–time spectrum $S_{0\tau}(\Omega) = S_0(\Omega) K_\tau(\Omega)$ if one assumes its FCC being independent of the target's properties and describes the radar with the function $K_0(\Omega)$.

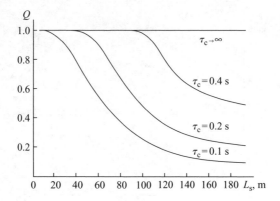

Figure 4.7   The parameter Q as a function of the synthesis range L_s at various signal correlation times τ_c

To conclude, the parameters of radar apertures for viewing fluctuating targets can be optimised by matching the characteristics $K_0(\Omega) \approx K_\tau(\Omega)$. The latter equality provides the imaging of a surface with nearly as much detail as possible potentially for a particular type of object. This equality can be obtained by choosing the value of L_s equal to $L_s = v\tau$, which means that the synthesis time should not be longer than the time of the signal correlation. As a result, the aperture resolution appears to be limited to $\delta_a = \lambda R / 2L_s$ but this choice of L_s provides the $(N = \Theta R/L_s)$ number of image realisations. The aperture contrast resolution, defined by the number of incoherent integrations N, is in turn independent of the signal coherence time τ. So the choice of $L_s > v\tau$ does not provide the desired spatial resolution but it decreases N, making the contrast resolution poorer.

The potentiality of the SAR in viewing low contrast targets can be conveniently described by the parameter $Q = Nd_h/(2\delta_a)$ equal to unity at zero fluctuations. If the fluctuations are present, Q essentially depends on the chosen synthesis range L_s (Fig. 4.7). For example, the signal fluctuations at $L_s < v\tau$ do not noticeably affect the image quality and $Q = 1$. At $L_s > v\tau$, the aperture performance proves to be inferior to its potentiality ($Q < 1$), since the real aperture resolution does not fit the chosen value of L_s but is rather defined by the signal correlation time τ.

We can draw the following conclusions from these results:

- To describe the imaging of fluctuating targets, one can make use of linear filtration theory, representing the radar as a filter with a certain FCC. The aperture can be considered as a device measuring the space–time spectrum of the object being viewed.
- One can suggest that the time fluctuations of the signal in the viewing channel create multiplicative noise decreasing the azimuthal resolution of the aperture.
- This approach provides a reasonable compromise between the potential azimuthal resolution and the aperture contrast resolution. This compromise can be achieved by choosing the synthesis time equal to the signal correlation time.

Imaging radars and partially coherent targets 99

The overall analysis of the results presented in this chapter shows that the available methods for describing the properties of sea surface images can be supplemented by a more general approach to SAR viewing of partially coherent objects. The concept of partial coherence allows one to cover a much larger class of targets and to describe the basic principles of their imaging. The advantages of this approach are as follows: first, it is based on a fairly general model of the radar signal. Expression (4.10) accounts for general and specific features of the viewing of fluctuating targets. We shall show in the following chapters that the correlation function of time fluctuations in Eq. (4.13) can be used, for example, to describe trajectory instabilities of the SAR carrier. Second, this approach provides an analytical description of the major statistical characteristics of images of partially coherent targets; these, in turn, enable one to evaluate image quality. Finally, the relative simplicity of mathematical calculations and the clear physical sense of the results obtained make this approach advantageous and convenient as a tool for solving practical tasks associated with SAR designing and for remote sensing of partially coherent targets.

Chapter 5

Radar systems for rotating target imaging (a holographic approach)

The possibility of using the rotation of an object to resolve its scattering centres was, probably, first shown by W. M. Brown and R. J. Fredericks [21]. Independently, microwave video imaging of rotating objects was demonstrated theoretically and experimentally by other researchers [109].

An analysis of three approaches (in terms of the antenna, range-Doppler and cross-correlation theories) was made in References 104 and 146 for the imaging of rotating targets. Here we discuss this problem in terms of a holographic approach.

5.1 Inverse synthesis of 1D microwave Fourier holograms

We shall start with the basic principles of inverse synthesis of microwave holograms of an object rotating around the centre of mass. The analysis will be based on the holographic approach discussed in Sections 1.2 and 2.4.

Lens-free optical Fourier holography [131] implies that an optical hologram is recorded when the amplitude and phase of the field scattered by the object are fixed in a certain range of bistatic angles $0 < \beta < \beta_0$ (Fig. 5.1). In the microwave range, this is equivalent to the displacement of the radar receiver along arc L of radius R_0 from point A to point B, while the transmitter remains immobile. A coherent background must be created by a reference supply located in the object plane. Since such a supply is unfeasible, the coherent background is created by an artificial reference wave in the radar receiver (Chapter 2). In further analysis, we shall use a model object made up of scattering centres described by Eq. (2.3). Then a direct synthesis along arc L of radius R_0 by a bistatic radar system (Fig. 5.1) can produce a classical microwave Fourier hologram [109], with a subsequent image reconstruction as a 1D distribution of the scattering centres and their effective scattering surfaces.

To discuss the principles of inverse synthesis and formation of a 1D microwave Fourier hologram, we shall make use of the well-known relation for uni- and bistatic

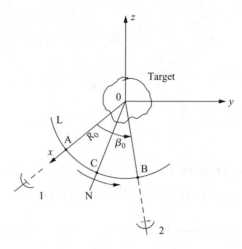

Figure 5.1 A schematic diagram of direct bistatic radar synthesis of a microwave hologram along arc L of a circle of radius R_0: 1 – transmitter, 2 – receiver

radars [69]. According to Kell's theorem, at small bistatic angles β the bistatic radar cross-section (RCS) for the angle α (Eq. 2.5) and the bistatic angle β is equal to the unistatic RCS measured along the bisectrix of the angle β at a frequency reduced by a factor of $\cos(\beta/2)$ (Chapter 2).

Kell's theorem and the fact that the rotation of a transmitter–receiver unit around the object can be replaced by the rotation of the object round its axis passing through the centre of mass normal to the radar viewing line lead one to the conclusion that such a unit, fixed at the point C (Fig. 5.2), can synthesise a 1D microwave Fourier hologram identical to a lens-free optical Fourier hologram. This approach was first discussed by S. A. Popov et al. [109].

In order to find analytical relations for the classical and synthesised Fourier holograms, let us consider the schematic diagram in Fig. 5.3. To simplify the calculations, we shall deal only with one kth scattering centre with the coordinates

$$r_{kx} = r_k \sin\theta_k \cos(\varphi + \varphi_k),$$

$$r_{ky} = r_k \sin\theta_k \sin(\varphi + \varphi_k), \qquad (5.1)$$

$$r_{kz} = r_k \cos\theta_k,$$

where $\varphi = \Omega t$ is the object rotation angle, $\Omega = |\vec{\Omega}|$ is the angular velocity of the rotating object, φ_k is the initial angle between the $\vec{r}_k$ vector projection on the xOz plane and the positive x-axis, and θ_k is the angle between the $\vec{r}_k$ vector and the positive z-axis. In our further analysis, we shall follow the References 109 and 145.

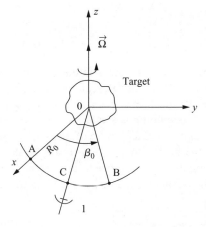

Figure 5.2 A schematic diagram of inverse synthesis of a microwave hologram by a unistatic radar located at point C

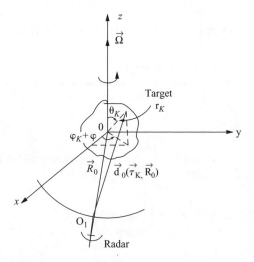

Figure 5.3 The geometry of data acquisition for the synthesis of a 1D microwave Fourier hologram of a rotating object

With Eq. (5.1), the input receiver signal can be described as a function of the object rotation angle:

$$\dot{u}_r(\varphi) = u_0 \sum_{k=1}^{N} \sigma_k \exp\left[-j\frac{4\pi}{\lambda_1}d(\vec{r}_k, \vec{R}_0) \cdot \exp i\omega_0 \frac{\varphi}{\Omega}\right], \qquad (5.2)$$

where

$$d(\vec{r}_k, \vec{R}_0) \cong R_0 \left\{ 1 - \frac{r_k}{R_0}[\sin\gamma \sin\theta \cos(\varphi + \theta_k) + \cos\gamma \cos\theta_k] \right\}, \quad (5.3)$$

$\lambda_1 = 2\pi c/\omega_0$ is the radar wavelength; σ_k is the amplitude coefficient accounting for the reflection characteristics of the kth scattering centre; $\gamma = \arctan(x_0/y_0)$ is the angle between the vector $\vec{R}_0$ and the positive z-axis; x_0, y_0, z_0 are the observation point coordinates; O_1 is the observation point; and $R_0 = |\vec{R}| = \sqrt{x_0^2 + z_0^2}$ is the distance between the observation point and the centre of mass of the object.

In order to derive the hologram function in a way shown in Chapter 2, it is reasonable to use the multiplication procedure performed by an amplitude-phase detector, followed by an averaging. The artificial reference signal is

$$\dot{u}_{\text{ref}}(\varphi) = u_0 \exp\left[-i\omega_0\left(\frac{\varphi}{\Omega}\right) + \psi\right], \quad (5.4)$$

where $t = \varphi/\Omega$ is the current moment of time and ψ is an arbitrary initial phase. Using Eqs (5.2) and (5.4), we can write down the hologram function in the form:

$$H(\varphi) = \langle \operatorname{Re} \dot{u}_r(\varphi) \operatorname{Re} \dot{u}_{\text{ref}}(\varphi) \rangle$$

$$= \frac{u_0^2}{2} \sum_{k=1}^{N} \sigma_k \cos\left\{ \frac{4\pi}{\lambda} r_k (\cos\gamma \cos\theta_k + \sin\gamma \sin\theta_k \cos(\varphi + \varphi_k)) \right\}, \quad (5.5)$$

where the sign $\langle \cdots \rangle$ stands for the averaging.

To derive Eq. (5.5), the arbitrary initial phase of the reference signal has been chosen such that

$$\frac{4\pi R_0}{\lambda_1} - \psi = 0.$$

By expanding the function $\cos(\varphi + \varphi_k)$ into the power series of φ and choosing only the first two terms of the series, we get

$$H(\varphi) = \frac{u_0^2}{2} \sum_{k=1}^{N} \sigma_k \cos 2\left[\beta_k - \frac{2\pi}{\lambda_1} r_k l_k(\varphi)\right], \quad (5.6)$$

with

$$\beta_k = \frac{2\pi}{\lambda_1} r_k (\cos\gamma \cos\theta_k + \sin\gamma \sin\theta_k \cos\varphi_k) \quad (5.7)$$

and

$$l_k(\varphi) = \sin\gamma \sin\theta_k \left(\varphi \sin\varphi_k + \frac{\varphi^2}{2}\cos\varphi_k - \frac{\varphi^3}{6}\sin\varphi_k\right). \quad (5.8)$$

Consider now the microwave hologram function of the same object (Fig. 5.1), obtained by a classical method. In this method, the radar receiver scans, with an angular velocity Ω, the surface of a cylinder of radius $R_0 \sin\gamma$, having the generatrix parallel to the z-axis. The transmitter is at the point A with the coordinates

$x_A = R_0 \sin\gamma$, $y_A = 0$, $z_A = R_0 \cos\gamma$, while the angle β_0 is equal to the rotation angle φ. Then the function $H_{\text{cl}}(\varphi)$ for the classical microwave Fourier hologram is

$$H_{\text{cl}}(\varphi) = \frac{u_0^2}{2} \sum_{k=1}^{N} \sigma_k \cos 2\left[\beta_k - \frac{\pi}{\lambda_1} r_k l_k(\varphi)\right], \tag{5.9}$$

where the functions β_k and $l_k(\varphi)$ are similar to those of Eqs (5.6) and (5.7).

A comparison of Eqs (5.6) and (5.9) shows that the function $H_{\text{cl}}(\varphi)$ differs from the function $H(\varphi)$ for the synthesised hologram of the same object in having the factor $(\frac{1}{2})$ in the second term of the argument $\cos 2[\cdots]$. It is clear that the synthesised hologram possesses a double capacity to change the argument and, hence, it has twice as high resolution because it looks like the classical hologram recorded in a field with a wavelength twice as short as the real one. This effect is due to the simultaneous scanning by several elements of the transmitter–object–receiver system. It is easy to see that a microwave hologram recorded by a simultaneous receiver–transmitter scanning of a fixed object along the arc L (Fig. 5.1) is totally identical to the $H_A(\varphi)$ hologram. In the case of inverse scanning, however, the rotation of the object alone is equivalent to the movement of two devices – the transmitter and the receiver.

We shall show below that the constant initial phase β_k does not affect the structure of microwave radar imagery. We shall use a simplified expression for the synthesised Fourier hologram:

$$H_1(\varphi) \cong \sum_{k=1}^{N} \sigma_k \cos\left[\frac{4\pi}{\lambda_1} r_k \sin\theta_k \cos(\varphi_k + \varphi)\right], \tag{5.10}$$

where r_k, θ_k, φ_k are the spherical coordinates of the kth centre. Equation (5.10) was derived from Eq. (5.5) on the assumption of $\gamma = 90°$ and is valid for the far-zone approximation.

Since the $H_1(\varphi)$ function basically coincides with $H_{\text{cl}}(\varphi)$, the image reconstruction from a synthesised Fourier hologram can be made in visible light, using the same techniques as those of optical Fourier holography [131].

Sometimes, a microwave hologram recorded on a flat transparency is placed in the front focal plane of the lens L (Fig. 5.4(a)). When the transparency is illuminated by a plane coherent light wave, two real conjugate images of the object, M and M', are formed near the rear focal plane of the lens. An alternative is to use a spherical transparency of radius F_0, illuminated by a coherent light beam converging at the sphere centre (Fig. 5.4(b)). The two variants are identical in the sense that the operations to be performed are the same. Practically, it is convenient to use the first variant but to analyse the second one.

If a microwave hologram is recorded on an optical transparency uniformly moving with velocity v_t, the angular coordinate $\alpha = v_t \tau / F_0$ on the transparency in the reconstruction space will be related to the angular coordinate $\varphi = \Omega\tau$ on the hologram in the recording space:

$$\alpha = \varphi v_e / \Omega F_0 = \varphi/\mu, \quad \mu = \Omega F_0 / v_t. \tag{5.11}$$

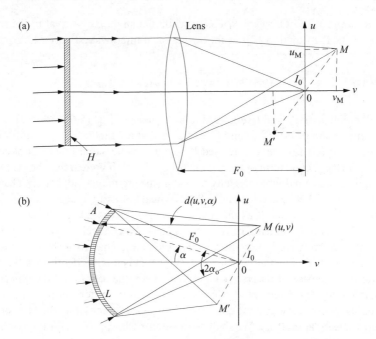

Figure 5.4 *Optical reconstruction of 1D microwave images from a quadrature Fourier hologram: (a) flat transparency, (b) spherical transparency*

For a hologram of a point object, the distribution of complex-valued light amplitudes in the image space u, v at the point $M(u_p, v_p)$ in the vicinity of the point O can be represented by an integral (at $\theta_k = 90°$):

$$E(u,v) = A \int_{-\alpha_0}^{\alpha_0} \left\{ 1 + \cos\left[2 \cdot \frac{2\pi}{\lambda_1} r_k \cos(\mu\alpha + \varphi_k) \right] \right\}$$

$$\times \exp\left[-j\frac{2\pi}{\lambda_2} d(u,v,\alpha) \right] d\alpha = I_0 + I_{+1} + I_{-1}, \qquad (5.12)$$

$$I_0 = A \int_{-\alpha_0}^{\alpha_0} \exp\left[-j\frac{2\pi}{\lambda_2} d(u,v,\alpha) \right] d\alpha,$$

$$I_{\pm 1} = \frac{A}{2} \int_{-\alpha_0}^{\alpha_0} \exp[j\psi_{\pm 1}(u,v,\alpha)] d\alpha,$$

$$I_{\pm 1}(u,v,\alpha) = \pm 4 \frac{\pi}{\lambda_1} r_k \cos(\mu\alpha + \varphi_k) - \frac{2\pi}{\lambda_2} d(u,v,\alpha),$$

$$d(u,v,\alpha) = [F_0^2 + 2F_0(v\cos\alpha - u\sin\alpha) + u^2 + v^2]^{1/2},$$

where λ_2 is the wavelength in the optical range; A is a complex-valued proportionality factor $A = (u_0^2/4)\sigma_1$; σ_1 is the amplitude coefficient accounting for the reflection characteristics of the scattering centre $k = 1$; $d(u, v, \alpha)$ is the distance between an arbitrary point in the arc L and the point M near the arc centre on the image; $2\alpha_0$ is the angular size of the hologram in the image space; and F_0 is the lens focal length.

The integrals I_0, I_{+1}, and I_{-1} describe the distribution of the complex-valued light amplitudes in the zeroth and first diffraction orders, both positive and negative. If the angular dimensions of the hologram are not too large, the functions $\cos(\mu\alpha + \varphi_k)$ and $d(u, v, \alpha)$ can be represented as the first term of the respective expansion series to write down the function $\psi(u, v, \alpha)$:

$$\psi_{\pm 1}(u, v, \alpha) = \frac{2\pi}{\lambda_2}\left[\alpha\left(2\frac{\lambda_2}{\lambda_1}\mu r_k \sin\varphi_k \pm u\right) + \frac{\alpha^2}{2}\left(2\frac{\lambda_2}{\lambda_1}\mu^2 r_k \cos\varphi_k \pm v\right)\right.$$
$$\left. - \frac{\alpha^3}{6}\left(2\frac{\lambda_2}{\lambda_1}\mu^3 r_k \sin\varphi_k \pm u\right)\right]. \tag{5.13}$$

Here we have omitted the constant expansion terms independent of the argument α. The coordinates of the points $M(u_M, v_M)$ and $M'(u'_M, v'_M)$, at which two conjugate images of the point object are formed, can be found from the expressions

$$\frac{\partial \psi(u, v, \alpha)}{\partial \alpha} = 0, \qquad \frac{\partial^2 \psi(u, v, \alpha)}{\partial \alpha^2} = 0. \tag{5.14}$$

With $x_M = r_k \sin\varphi_k$ and $y_M = r_k \cos\varphi_k$, using Eq. (5.14), we get

$$u_{M,M'} = \pm 2\mu\frac{\lambda_2}{\lambda_1}x_M, \qquad v_{M,M'} = \pm 2\mu^2\frac{\lambda_2}{\lambda_1}y_M. \tag{5.15}$$

Equation (5.15), in turn, gives the transverse and longitudinal scales of the image being reconstructed:

$$m_y = \left|\frac{v_M}{y_M}\right| = 2\mu^2\frac{\lambda_2}{\lambda_1}, \qquad m_x = \left|\frac{u_M}{x_M}\right| = 2\mu\frac{\lambda_2}{\lambda_1}. \tag{5.16}$$

An undistorted image of an object can be reconstructed only if all the derivatives of $\psi(u, v, \alpha)$ are simultaneously equal to zero with respect to the argument α. It is easy to show that this condition is met at one point (M and M') at $\mu = \Omega F_0/v_t$ or $\varphi \equiv \alpha$. The latter identity defines the criterion for optical processing of synthesised Fourier holograms: the aperture angles in the recording and reconstruction space must be the same. If the reconstruction procedure has been designed in the optimal way, we have $m_x = m_y = m$, and the object is reproduced without distortions along the longitudinal and transverse directions.

A specific feature of a synthesised Fourier hologram is that the resolution obtained is independent of the distance to the object. Indeed, let us take the following expression to be the measure of the resolving power:

$$\Delta = |I(u_M)|^{-2}\int_{-\infty}^{\infty}|I(u)|^2 du, \tag{5.17}$$

where $|I(u)|^2$ is the light intensity distribution across the scattering centre image and u_M is the coordinate of the maximum intensity of the image focusing.

Equation (5.17) describes the receiver pulse response to the point object. Then, neglecting all the terms in Eq. (5.13) except for the first one and using the scale relations of Eq. (5.16), we can define the resolving power of the object as

$$\Delta_x(\lambda_1, \psi_S) = \frac{\Delta u}{m_x} = \frac{\lambda_1}{4\varphi_0} = \frac{\lambda_1}{2\psi_S}, \quad (5.18)$$

where ψ_S is the object angle variation during the recording. Therefore, when the hologram angles are small, the resolving power of the object varies with the wavelength and the synthesised aperture angles, rather than with the distance to the object or the reconstruction parameters.

With the scale relations from Eq. (5.16), we find for $\mu = 1$

$$m_y = m_x = 2\frac{\lambda_2}{\lambda_1}.$$

Then the criterion described by Eq. (5.18) can yield the resolution of a video microwave image:

$$\Delta_u(\alpha_0) = \Delta_x(\lambda_1, \psi_S) m_x = \frac{\lambda_2}{2\varphi_0}. \quad (5.19)$$

It follows from Eq. (5.19) that the resolution of a microwave image obtained by inverse synthesis and optimal processing is fully consistent with the Abbe criterion for optical devices (Chapter 1).

Consider now distortions arising from the reconstruction of a microwave image. These are defined by the high-order terms of Eq. (5.13) for the following reason. When an image is viewed in one plane, some of the scattering centres are shifted relative to this plane, that is, they are defocused. With the quadratic term of Eq. (5.13), the field distribution in a defocused point image is defined as

$$I_{+1}(p, t_0) = A\frac{\alpha_0}{t_0} \exp\left[\pi j \left(\frac{4r_x}{\lambda_1} - \frac{p^2}{2}\right)\right]$$
$$\times \{C(t_0 + p) + C(t_0 - p) + j[S(t_0 + p) + S(t - p)]\}, \quad (5.20)$$

where $t = \sqrt{2(v_M - v)/\lambda_2}$ describes the viewing plane shift relative to the focusing plane and $p = u_M/\lambda_2 t$, $t_0 = \alpha_0 t$ and $S(z)$, $C(z)$ are the Fresnel integrals.

The resolution of a defocused microwave image is described by the function

$$\Delta(t_o) = \int_{-\infty}^{\infty} \frac{|I_{+1}(p, t_0)|^2}{|I_{+1}(O, t_0)|^2} dp \quad (5.21)$$

shown in Fig. 5.5. Obviously, the best resolution $\hat{\Delta} = 1.2$ is achieved at a certain optimal value of $t_0 = \hat{t}_0 = 1$ and an optimal aperture size

$$\hat{\alpha}_0 = [2(v_M - v)/\lambda_2]^{-1/2}. \quad (5.22)$$

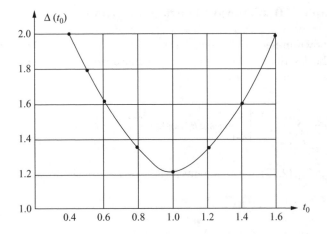

Figure 5.5 *The dependence of microwave image resolution on the normalised aperture angle of the hologram*

At $v = 0$, when the viewing plane is superimposed with the focal plane of the lens, we can use Eq. (5.15) to get

$$\hat{\alpha}_0 = \left(2\mu\sqrt{y_M/\lambda_1}\right)^{-1} = \left(\mu\sqrt{\tau_{max}}\right)^{-1}, \qquad (5.23)$$

where $\tau_{max} = 2L_{max}/\lambda_1$ is the maximum longitudinal dimension of the object, expressed as half-wavelengths.

As the size of the object or the aperture increases, the influence of the high-order terms of Eq. (5.13) becomes more pronounced resulting in distortions and a lower resolution. These factors impose constraints on the synthesised aperture size.

The image reconstruction of microwave Fourier holograms has some specificity associated with the way the artificial reference wave is created. If the reference signal phase is not modulated, the phase of the coherent reference background along the hologram is constant, a situation equivalent to the position of a point object at the rotation centre. So during the reconstruction, the three images – that of the reference source and the two conjugate images of the object – overlap. To separate these images, one should introduce a space carrier frequency (SCF) by changing the phase of the reference signal at a constant rate, like in the expression

$$d\psi/d\tau \geq 4\pi \Omega r_{max}/\lambda_1, \qquad (5.24)$$

where r_{max} is the vector radius of the scattering centre located at the maximum distance from the object rotation centre.

The reference wave phase can be modulated by a phase shifter or by introducing translational motion along the viewing line, in addition to the rotational motion. In the latter case, the translational velocity v must satisfy the inequality $v > \Omega r_{max}$.

5.2 Complex 1D microwave Fourier holograms

We have shown in Section 5.1 that a 1D quadrature microwave Fourier hologram $H_1(\varphi)$ can be described by Eq. (5.10). A conjugate quadrature Fourier hologram with a $\pi/2$ phase shift has the form:

$$H_2(\varphi) \cong \sum_{k=1}^{N} \sigma_k \sin\left[\frac{4\pi}{\lambda_1} r_k \sin\theta_k \cos(\varphi_k + \varphi)\right]. \tag{5.25}$$

According to Eq. (2.23), the holograms $H_1(\varphi)$ and $H_2(\varphi)$ can form a complex Fourier hologram:

$$H(\varphi) = H_1(\varphi) + jH_2(\varphi) = \sum_{k=1}^{N} \sigma_k \exp\left[j\frac{4\pi}{\lambda_1} r_k \sin\theta_k \cos(\varphi_k + \varphi)\right]. \tag{5.26}$$

This expression can be re-written in a simpler form:

$$H(x) = u\exp(j\Phi), \tag{5.27}$$

where u and Φ are the amplitude and phase (in the recording plane) of the total field scattered by the object. The argument φ of the H function has been replaced by the linear x-coordinate, since a 1D microwave hologram is recorded on a flat transparency.

The image reconstruction by a plane wave in a paraxial approximation is reduced to the Fourier transformation of the hologram function, assuming for simplicity that the recording and the reconstruction are performed at the same wavelength:

$$V(\omega_x) = \int_{-\infty}^{\infty} H(x)\exp(-j\omega_x x)\,dx, \tag{5.28}$$

where ω_x is the space frequency corresponding to the coordinate in the image plane.

The substitution into Eq. (5.28) of the expressions for the quadrature holograms in Eqs (5.10) and (5.25), re-written as Eq. (5.27), gives

$$V_1(\omega_x) = \frac{1}{2}\left[\int_{-\infty}^{\infty} u\exp(j\Phi)\exp(-j\omega_x x)\,dx \right.$$

$$\left. + \int_{-\infty}^{\infty} u\exp(-j\Phi)\exp(-j\omega_x x)\,dx\right], \tag{5.29}$$

$$V_2(\omega_x) = \frac{1}{2j}\left[\int_{-\infty}^{\infty} u \exp(j\Phi)\exp(-j\omega_x x)\,dx\right.$$
$$\left. - \int_{-\infty}^{\infty} u \exp(-j\Phi)\exp(-j\omega_x x)\,dx\right]. \tag{5.30}$$

It is seen that each quadrature hologram gives two conjugate images described by the appropriate terms in Eqs (5.29) and (5.30).

In a complex hologram, the first quadrature component gives two conjugate images in Eq. (5.29), while the second component reconstructs the images

$$V_2(\omega_x) = \frac{1}{2}\left[\int_{-\infty}^{\infty} u \exp(j\Phi)\exp(-j\omega_x x)\,dx\right.$$
$$\left. - \int_{-\infty}^{\infty} u \exp(-j\Phi)\exp(-j\omega_x x)\,dx\right]. \tag{5.31}$$

The first terms in Eqs (5.29) and (5.31) are identical, while the second terms differ in the phase by the value π. A combined reconstruction after summing up the fields in Eqs (5.29) and (5.31) yields one pair of conjugate images that enhance each other and another pair of images that annihilate each other; so we eventually have

$$V(\omega_x) = \int_{-\infty}^{\infty} u \exp(j\Phi)\exp(-j\omega_x x)\,dx. \tag{5.32}$$

The complex-valued function $V(\omega_x)$ describes the only image reconstructed from a complex hologram [145]. The image intensity can be defined as

$$W(\omega_x) = |V(\omega_x)|^2. \tag{5.33}$$

To illustrate, consider the case when the object is a point and the parameters θ_1 and φ_1 are equal to $\pi/2$. For small values of φ ($\varphi < 1$ rad.) and $\varphi = \Omega x/v_t$, where v_t is the velocity of the recording transparency, Eq. (5.26) reduces to

$$H(x) \cong u \exp\left(j\frac{4\pi}{\lambda_1} r\frac{\Omega}{v_t} x\right). \tag{5.34}$$

Since the hologram is recorded in a finite time interval, $\tau \in [-T/2, T/2]$, Eq. (5.28) yields

$$V(\omega_x) = \int_{-v_t T/2}^{v_t T/2} H(x)\exp(-j\omega_x x)\,dx. \tag{5.35}$$

The substitution of Eq. (5.34) into Eq. (5.35) and the integration give

$$V(\omega_x) = 2\sigma \sin\left[\left(\frac{4\pi}{\lambda_1}r\frac{\Omega}{v_t} - \omega_x\right)\frac{v_t T}{2}\right] / \left(\frac{4\pi}{\lambda_1}r\frac{\Omega}{v_t} - \omega_x\right). \quad (5.36)$$

Clearly, this function is of the $\sin z/z$ type and has a maximum at $\omega_x = (4\pi/\lambda_1)r(\Omega/v_t)$, which corresponds to the image of the point.

Digital reconstruction reduces to the calculation of the integral in Eq. (5.28) and has no zeroth order. So a complex hologram can be formed without introducing the carrier frequency, which decreases the amount of data to be processed: a single quadrature hologram requires, at least, twice as many discrete counts because of the high carrier frequency.

Optical reconstruction produces the zeroth order, in addition to a single image, because of the presence of the reference level of H_r (Eq. (2.20)). During the processing of a complex hologram recorded without the carrier frequency, the zeroth order overlaps the image. Their spatial separation can be made by just introducing the carrier frequency. Then the use of a complex hologram has no sense, since one does not have to remove the conjugate image. Besides, the optical reconstruction of a complex hologram is hard to make due to the strict requirements on the adjustment of the two-channel processing suggested in Reference 35. Thus, complex microwave holograms should be recorded without introducing the carrier frequency and reconstructed only digitally.

5.3 Simulation of microwave Fourier holograms

A comparison of various techniques applied in microwave Fourier holography can be made using a special algorithm for digital simulation of 1D quadrature and complex hologram recording and reconstruction for simple objects. The algorithm consists of two units, one of which records a hologram following Eq. (5.26) and the other reconstructs the image, that is, calculates the integral of Eq. (5.28). The image reconstruction from individual quadrature holograms is performed using an additional procedure for the calculation of the Fourier integrals of the real functions H_1 and H_2 from the Fourier transform of the complex function $H = H_1 + jH_2$.

Figure 5.6(a–c) illustrates some of the results of the digital simulation. The ordinate shows the image intensity in relative units and the abscissa the image size. In digital reconstruction, a microwave image represents a series of discrete counts spaced at a distance $\lambda_1/2\psi_S$. The model object consisted of two scattering centres arranged to form a dumb-bell structure of $10\lambda_1$ in length, which rotated at a constant angular velocity round the centre of mass. The quantity θ_k (Fig. 5.3) was taken to be equal to $\pi/2$. The image illustrated in Fig. 5.6(a) was reconstructed from a single quadrature hologram. Peaks 1 and 2 correspond to one conjugate image of the two scattering centres and peaks 3 and 4 to the other. The image separation was made using the SCF, whose introduction was simulated by the radial displacement (with the velocity v_l) of the object rotation centre relative to the receiver. One of the conjugate images vanished during the processing of the complex hologram (Fig. 5(b)), so the carrier frequency

Radar systems for rotating target imaging (holographic approach) 113

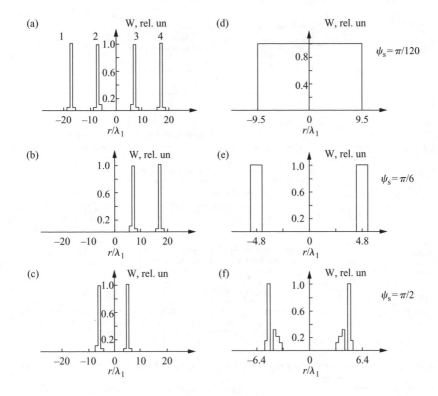

Figure 5.6 Microwave images reconstructed from Fourier holograms: (a) quadrature hologram, (b) complex hologram with carrier frequency, (c) complex hologram without carrier frequency and (d,e,f) the variation of the reconstructed image with the hologram angle ψ_s (complex hologram without carrier frequency)

was not needed. This is clearly seen in Fig. 5.6(c) showing the image reconstructed from a complex hologram recorded without the carrier frequency.

Figure 5.6(d–f) presents the variation of the reconstructed image with the hologram angle. The comparison of these results supports the above conclusion that there is an optimal size of the synthesised aperture. As the angle ψ_S becomes larger, the resolution increases to a certain limit, beyond which distortions arise in the image structure. The resolving power of this technique estimated from the results of the digital simulation is $\sim \lambda_1$.

Currently, there are two methods used in microwave Fourier holography. One is based on the recording of a single quadrature phase-amplitude hologram of the type described by Eq. (5.10) with the carrier frequency and optical image reconstruction. The other method records a complex hologram of the type described by Eq. (5.26) without introducing the carrier frequency but using a digital image reconstruction.

The application of the first method involves some problems associated with the use of an anechoic chamber (AEC), because the linear displacement of the object for introducing the carrier frequency leads to the camera decompensation. So we

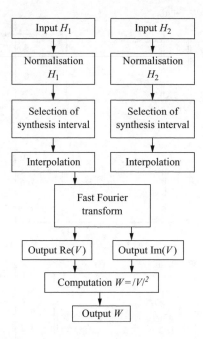

Figure 5.7 The algorithm of digital processing of 1D microwave complex Fourier holograms

recommend the second technique when one uses an anechoic camera. We shall discuss some of the results obtained by the second method.

Figure 5.7 illustrates the algorithm of digital image reconstruction, which operates as follows. The setting of discrete data is followed by their normalisation, that is, the data are reduced to the variation range $[-1, 1]$. The hologram is usually recorded for a full 2π rad rotation; so for the subsequent processing, one selects a series of counts in such a way that their number describes the optimal aperture and their position in the array corresponds to the required aspect. An interpolation unit makes it possible to reduce the number of signal records to 2^m, where m is a natural number. The image reconstruction is performed by a Fourier transform unit using the FFT algorithm for the complex-valued function $H(x)$. Arrays of $\text{Re}(V)$ and $\text{Im}(V)$ numbers that define the image, whose intensity is found as $W = \text{Re}^2(V) + \text{Im}^2(V)$, are produced at the unit output.

Figure 5.8 presents the results of digital processing of 1D complex Fourier holograms recorded experimentally with an anechoic camera. The image intensity is plotted in relative units along the y-axis and its linear dimension along the x-axis. The object is a metallic sphere of radius $0.3\lambda_1$, rotating along a circumference of radius $3\lambda_1$. The positions of the point image in Fig. 5.8(a–c) are different and vary with the object aspect ψ_0 as shown schematically in each figure.

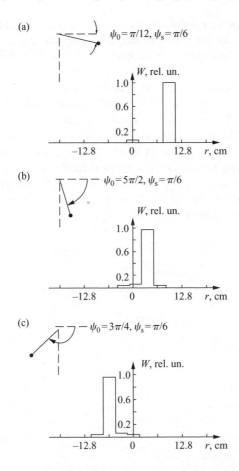

Figure 5.8 A microwave image of a point object, reconstructed digitally from a complex Fourier hologram as a function of the object's aspects $\Psi_0(\Psi_s = \pi/6)$: (a) $\Psi_0 = \pi/12$, (b) $\Psi_0 = 5\pi/2$ and (c) $\Psi_0 = 3\pi/4$.

The methods we have discussed have some advantages and limitations. The recording of single quadrature holograms is made in one channel but requires that the carrier frequency should be introduced in this way or another. The recording of complex holograms does not require the carrier frequency but it is more complicated because the channels must have a strict quadrature character, their parameters must be identical, and the measurements must be well synchronised. However, the recording errors associated with these characteristics of a two-channel system can be easily eliminated by the processing. (We have mentioned above that complex microwave Fourier holograms should be processed only digitally.) The image reconstruction from quadrature holograms can be made both digitally and optically. The possibility of recording a hologram in a form suitable for digital processing increases the dynamic range of the system. It does not then need the use of sophisticated units, such as high-resolution cathode-ray tubes or high-precision focusing and deflecting devices. In optical processing, the aperture size is normally limited by the characteristics of the reconstruction unit, so it cannot be made optimal. On the other hand, optical processing allows re-focusing of the observation plane without difficulty, providing a 2D image (in longitudinal and transversal directions).

The investigation and analysis of methods for microwave Fourier holography have shown that they can be successfully used for imaging objects which can be represented as an array of scattering centres. These methods are of interest to those studying diffraction with anechoic cameras (Chapter 9), in particular, for the experimental verification of the applicability of the physical theory of diffraction developed by P. Ya. Ufimtzev [137] and of the geometrical theory of diffraction by J. B. Keller [70]. These methods can also be useful in designing radar systems with an inversely synthesised aperture (Chapter 9).

Chapter 6

Radar systems for rotating target imaging (a tomographic approach)

6.1 Processing in frequency and space domains

Section 2.4.2 discussed the tomographic approach to target imaging in two-dimensional (2D) viewing geometry. We suggested an algorithm for processing in the frequency domain, which finds the reflectivity function $\hat{g}(x,y)$ from Eq. (2.48).

The first procedure to be performed is to reconstruct an image in the frequency domain by calculating the N number of discrete Fourier transform (DFT) records of the echo complex envelope

$$P_\theta(l,m) = \sum_{n=0}^{N=1} s_v(n\Delta t, m\delta\theta) \exp(-j2\pi \ln/N) \tag{6.1}$$

for each of the M number of the target angular positions $m\delta\theta$, $m = 0, \ldots, M-1$. The pixels found in this way are located at the polar grid nodes formed by the interceptions of concentric circumferences separated by the frequency step $1/N\Delta t$ and rotated by the radial beam angle $\delta\theta$ from one another.

Since an inverse DFT can be made only on a rectangular grid, the second procedure should include the finding of pixels at the equidistant nodes of a rectangular grid, using the $P_\theta(l,m)$ values obtained by the first procedure. This is followed by a 2D inverse DFT computation of the target reflectivity $\hat{g}(x_i, y_j)$ at the rectangular grid nodes.

This algorithm has two important features that deserve attention. First, since the complex envelope of an echo signal is finite, there are distortions near the $\pm 1/2\Delta t$ boundaries of the major period of the $P_\theta(l,m)$ spectrum. The distortions arise from the superposition of high-frequency components of the adjacent spectral periods. Besides, the high-frequency spectrum may contain noise that dominates over the signal data. To reduce the noise, one has to resort to weighting by multiplying the $P_\theta(l,m)$ DFT data by a 'window' function. The choice of such a function should be based on the consideration of how much the noise abates the radar data and what kind of target is being probed [57].

Second, since the radar is a coherent system, it seems important to define the discretisation step $\delta\theta$ of the θ angle as the target aspect changes. The criterion for choosing a $\delta\theta$ value can be formulated as follows: the phase shift of the echo signal from the point scatterer most remote from the target centre of mass should not be larger than π when the target aspect changes by $\delta\theta$. This criterion is written as

$$\delta\theta \leq \lambda_c/4|\bar{r}_o|_{max}. \tag{6.2}$$

This expression is valid for relatively narrowband signals, whose spectral width is much less than the carrier frequency. Otherwise, one should substitute λ_c in Eq. (6.2) by the wavelength of the highest frequency component in the signal spectrum.

It is worth noting that the method of synthesising the so-called unfocused aperture is a particular case of the above processing algorithm for the frequency domain. The movement of a point scatterer along an arc is approximated by the movement along a tangent to it. By substituting $v = y\cos\theta - x\sin\theta$ into Eq. (2.40) and using $\sin\theta \approx \theta$ and $\cos\theta \approx 1 - \theta^2/2$, we get

$$S(f) = H(f) \iint\limits_{-\infty}^{\infty} g(x,y)\{\exp[j(k_c+k)\theta_y^2]\}$$

$$\times \exp[-j2(k_c+k)y + j2(k_c+k)\theta x]\,dx\,dy.$$

If we eliminate the squared phase term, it will be clear that the $\hat{g}(x, y)$ function can be reconstructed by an inverse Fourier transform (IFT) over the rectangular raster which has replaced the respective region of the polar raster. This approximation works well only if the aspect variation during the data acquisition was small.

Let us discuss now the processing algorithm for the space domain, or the convolution algorithm. For this, Eq. (2.48) will be transformed from the Cartesian to polar coordinates:

$$\hat{g}(x,y) = \int_0^\pi d\theta \int_{-\infty}^{\infty} S_\theta(f_p)|f_p|\exp[j2\pi f_p r \cos(\theta-\varphi)]\,df_p. \tag{6.3}$$

The inner integral in Eq. (6.3) represents the IFT of the product of f_p and the function defined by expression (2.43). The result is the convolution of the quantity $F^{-1}\{S_\theta(f_p)\}$ with the so-called kernel function $q(v) = F^{-1}\{|f_p|\}$. If one uses the window function $F(f_p)$ to reduce the effect of high-frequency spectral noise, one gets

$$g(v) = F^{-1}\{|f_p|F(f_p)\}. \tag{6.4}$$

The result of the integration with respect to the variable f_p in Eq. (6.3) using Eq. (6.4) is known as a convolutional projection. It can be used for making a back projection procedure:

$$\hat{g}(x,y) = \int_0^\pi \xi_\theta[r\cos(\theta-\varphi)]\,d\theta. \tag{6.5}$$

Radar systems for rotating target imaging (tomographic approach)

This procedure implies the integration of the contribution of each convolutional projection $\xi_0(\cdot)$ to the resulting image. The substitution of the integral in Eq. (6.4) by the Riman sum gives

$$\hat{g}(x_i, y_j) = \sum_{m=0}^{M=1} \xi_0[r(x_i, y_j)m\theta]\,\delta\theta, \qquad (6.6)$$

where

$$r(x_i, y_j, m\delta\theta) = \sqrt{x_i^2 + y_j^2}\cos[m\delta\theta - \text{arctg}(x_i/y_j)]. \qquad (6.7)$$

The latter expression is used to find (by interpolation) the contribution of the convolutional projection obtained at the mth target aspect to each of the (x_i, y_j) pixels of the rectangular image grid.

An important advantage of the convolution algorithm is the possibility of processing data as they become handy, because the contribution of every projection to the final image is computed individually.

If the transmitter signal contains a finite number L of discrete frequencies, Eq. (6.3) will take the form:

$$\hat{g}(x, y) = \sum_{l=1}^{L} \frac{4\pi f_p l}{c} \int_0^\pi S_\theta(f_p l) \exp[j 2\pi f_p l r \cos(\theta - \varphi)]\,d\theta \qquad (6.8)$$

and the processing algorithm reduces to summing up 1D integrals with respect to the variable θ. We can make computations with formula (6.8) in two ways. One is to calculate the integral for every value of (x_i, y_j) and the other is to solve the subintegral expression for the M number of aspects for every frequency value, followed by interpolation, as in the common convolution algorithm.

Thus, radar imaging of extended compact targets by inverse aperture synthesis can be made by using a number of algorithms well known in computerised tomography. The application of the convolution algorithm of the back projection method allows a reduction in the imaging time, as compared with the time of reconstruction in the frequency domain, due to the processing of individual echo signals. The interpolation can be omitted in the case of discrete-frequency transmitter signals, giving an additional reduction in the processing time.

Another important feature of an imaging radar is its coherence, so it provides more information than conventional systems using computerised tomography. On the other hand, coherence must be maintained in all of the radar units during the operation. This circumstance also imposes restrictions on the minimum repetition rate of transmitter pulses.

6.2 Processing in 3D viewing geometry: 2D and 3D imaging

It has been shown in Chapter 5 that inverse aperture synthesis is the most promising technique for imaging extended proper and extended compact targets with a high angular resolution. The fact that such targets can be imaged during their arbitrary motion makes it possible to use this technique in available radar systems (Chapter 9).

The conditions for microwave hologram recording are primarily determined by the application of the images to be obtained. For example, if radar responses are studied in an anechoic chamber (AEC) (Chapter 9), it is sufficient to use a 2D geometry with an equidistant arrangement of the aspect angles. The target rotates uniformly around the axis normal to the line of sight. By deviating the rotation axis from this normal after every measurement run, one can, in principle, obtain 2D images even with monochromatic radar pulses.

6.2.1 The conditions for hologram recording

There are a number of applied tasks when the target aspect variation must reflect natural viewing conditions. Let us consider the aspect variation relative to the line of sight of a ground radar viewing a hypothetical satellite moving at an altitude $H = 400$ km along a circular orbit with the inclination $i = 97°$ (Fig. 6.1). The target is assumed to be perfectly stabilised in the orbital coordinates, and its aspect in the orbital plane is defined by the angle α between the longitudinal construction line and the projection of the line of sight onto the orbital plane. The angle β between the line of sight and the orbital plane describes the aspect variation in the plane normal to orbital plane. The analysis of the plots presented shows that the aspect variation of this class of targets during hologram recording in real viewing conditions should be characterised by (1) a 3D viewing geometry and (2) a non-equidistant arrangement of samples within the view zone.

To derive analytical relations for the description of a microwave hologram for 3D viewing geometry, we shall consider the following conditions for viewing an orbiting satellite. The target is scanned by a ground coherent radar transmitting a probing signal with the carrier frequency f_0 and the modulation function $w(t)$ from Eq. (2.30). The radar measures the amplitude and phase of the echo signal (for a narrowband signal $\dot{w}(t) = A$, where A is the complex envelope amplitude).

The target is large relative to the wavelength λ of the radar carrier oscillation, such that the target can be represented as an ensemble of individual and independent scatterers. Every scatterer is rigidly bound to the target's centre of mass or moves across its surface as its aspect changes with respect to the radar. The position of the nth scatterer at any moment of time is defined by the radius vector $\vec{r}_{no}$ with the origin at point O rigidly bound to the target's centre of mass.

The positions of the arbitrary nth scatterer and the rotation centre of the satellite will be described by the radius vectors $\vec{r}_{no}$ and $\vec{R}_o$, respectively (Fig. 2.8). In the general case of 3D viewing geometry, an echo signal is defined, within the accuracy of a constant factor, as

$$S_v(t) = \int_V g(\vec{r}_{no})\{w(t - 2|\vec{R}_o|/c - 2\hat{r}_n/c)\exp(-j2\pi f_0 2|\vec{R}_o|/c)\}$$

$$\times \exp(-j2\pi f_0 2\hat{r}_n/c)\,d\vec{r}_{no}. \tag{6.9}$$

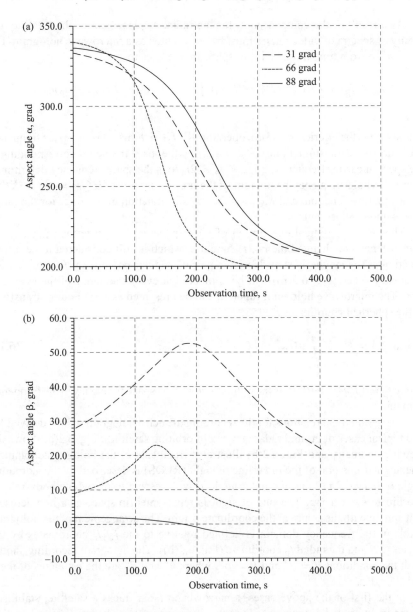

Figure 6.1 The aspect variation relative to the line of sight of a ground radar as a function of the viewing time for a satellite at the culmination altitudes of 31°, 66° and 88°: (a) aspect α and (b) aspect β

It follows from Eq. (2.34) that signal noise due to the presence of coordinate information can be corrected by the receiver. The correction consists in selecting the time strobe position in accordance with the delay $2|\vec{R}_o|/c$ and in introducing the phase factor $\exp\lfloor j2\pi f_0(2|\vec{R}_o|/c)\rfloor$ in the reference signal during the coherent sensing.

As a result of the compensation for the radial displacement of the satellite, the family of spectra of video pulses must be represented as a microwave hologram. For this, we go from time frequencies to space frequencies to get

$$S(f_{po}+f_p) = F\{S_v(ct/2)\} = H(f_p)\int_v g(\vec{r}_{no})\exp[-j2\pi(f_{po}+f_p)\vec{d}(t)]\,d\vec{r}_{no}, \quad (6.10)$$

where $F\{\cdot\}$ is the Fourier transform operator, $W(f_p) = F\{w(v)\}$ is the space frequency spectrum of the transmitter pulse, $f_{po} = 2f_o/c$ is the space frequency corresponding to the spectral carrier frequency, $2f_l/c < f_p < 2f_u/c$ is the space frequency determined over the whole frequency bandwidth of the transmitter pulse, $H(f_p) = W(f_p)K(f_p)$ is the aperture function, and $K(f_p)$ is the transfer function of the filter for the range processing of video pulses.

The above analytical description of video pulse spectra in terms of space frequencies has not changed the $\hat{r}_{no}(t)$ function, which is still considered to be a time function at the synthesis step. Now the pair of angular coordinates θ, B in the 3D frequency space (Fig. 6.2(b)) will be compared at every moment t of the synthesis step. The microwave hologram function can be presented as a 3D Fourier transform in the spherical coordinates f_p, θ, B:

$$S(\vec{f}_p) = H(\vec{f}_p)\int_v g(\vec{r}_{no})\exp(-j2\pi\vec{f}_p\vec{r}_{no})\,d\vec{r}_{no}, \quad (6.11)$$

where $\vec{f}_p = (f_{po}+f_p)\vec{e}(\theta, B)$ is the radius vector of the space frequency in the frequency domain.

The geometrical relations for the recording of such a hologram will be derived for two typical cases of ground radar viewing of orbiting satellites. Fig. 6.2(a) shows the viewing geometry and Fig. 6.2(b) illustrates fragments of the holograms obtained. The angular position of the radar line of sight (RLOS) is described by the azimuthal angle $\theta = \alpha - 3\pi/2$ and the polar angle β with respect to the whole body-related coordinate system xyz. The line of sight is represented in space as a line across a unit sphere with the centre at the coordinate origin. The arrangement of the hologram pixels in the frequency domain is defined relative to the $f_x f_y f_z$ coordinates by the angles θ, B and the radial f_p coordinate (Fig. 6.2(b)). The hologram recording should meet the conditions $\theta = \theta^*$ and $B = \beta^*$, where θ^* and β^* are the estimates of the θ and β angles.

In the first of the above cases, a narrowband radar tracks a satellite, stabilised by the body-related coordinates along the three axes, during its translational motion along the orbit. The line of sight turns relative to the satellite to describe a curve on the unit sphere (the left side of Fig. 6.2(a)), which represents an arc in the xy plane if the radar is located in the orbital plane, or a 3D curve in all other cases. If the radar transmits a continuous wave, the hologram reproduces the shape of this line on the sphere f_{po} in the frequency domain (Fig. 6.2(b)).

If a radar transmits a pulsed signal with the repetition rate F_r or if a continuous echo signal is appropriately discretised, a hologram will represent a series of individual

Radar systems for rotating target imaging (tomographic approach) 123

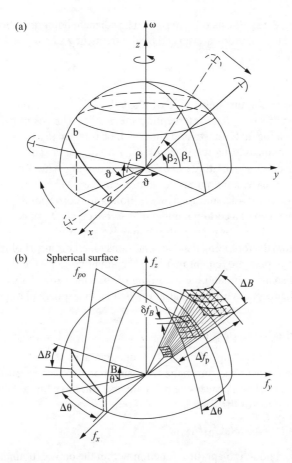

Figure 6.2 *Geometrical relations for 3D microwave hologram recording: (a) data acquisition geometry; a–b, trajectory projection onto a unit surface relative to the radar motion and (b) hologram recording geometry*

samples separated by $\delta f_\psi = f_{po}\dot{\theta}^* \cos\beta^*/F_r$, where $\dot{\theta}^* = d\dot{\theta}^*(t)/dt$ is the angular velocity of the satellite rotation in the orbital plane.

In the second case, one gets a wideband hologram of a satellite stabilised by rotation of the body-related coordinates around the z-axis (the right side of Fig. 6.2(b)). During the tracking, the angle between the line of sight and the rotation axis changes slowly by the value $\Delta\beta = \beta_2 - \beta_1$ with $\dot{\beta} \ll \dot{\theta}$. The interception of the unit sphere surface by the line of sight forms a spiral confined between two conic surfaces with the half angles $\pi/2 - \beta_1$ and $\pi/2 - \beta_2$ at the vertex. The resulting hologram represents a multiplicity of real beams that form a spiral band (Fig. 6.2(b)). The band is transversely bounded by two spherical surfaces and is 'fitted' between two conical surfaces, with $B_1 = \beta_1$ and $B_2 = \beta_2$. The radii of the spheres are equal to the lower f_{pl} and upper f_{pu} space frequencies of the hologram. Figure 6.2(b) shows a fragment of such a hologram bounded by the azimuthal step $\Delta\theta$, while the satellite makes the $\dot{\theta}\Delta t/2\pi$ number of rotations during the synthesis time step Δt. The adjacent hologram slices synthesised during consecutive rotations are spaced by the frequency step $\delta f_u = 2\pi f_{po} \dot{\beta}^* / \dot{\theta}^*$.

Under the condition

$$\delta f_u^{-1} \geq D,$$

where D is the maximum linear size of a satellite, the resolution can be achieved by the synthesis in the plane intercepting the z-axis. The resulting 3D wideband hologram containing, at least, several slices will be referred to as a surface hologram. A surface hologram is usually synthesised by a wideband radar, when tracking a satellite stabilised along the three axes, or when dealing with a model target in an AEC. In the latter case, a hologram lies entirely in the f_x–f_y plane.

Every beam of a wideband microwave hologram corresponds to a single echo signal and is made up of a certain number of discrete pixels, L, since digital hologram processing implies discretisation of the echo pulse spectrum.

It is clear from the foregoing that the conditions for recording a hologram of a target performing a complex movement relative to an imaging radar are the compensation for its radial displacement and the recording of the video signal spectrum in a form adequate for the respective aspect variation, that is, in a spherical or polar geometry.

6.2.2 Preprocessing of radar data

The preliminary processing of radar data integrated in the form of a microwave hologram to be further used for image reconstruction can be described in terms of a linear filtering model as a processing by an inverse filter in a limited frequency band. The transfer function of the filter is

$$H_f(\vec{f}_p) = H^{-1}(\vec{f}_p) H_o(\vec{f}_p) H_r(\vec{f}_p), \qquad (6.12)$$

where $H_r(\vec{f}_p)$ is a non-zero aperture function within the chosen boundaries $\vec{f}_{ph}$ of the hologram (Fig. 6.2(b)):

$$H_o(\vec{f}_p) = \text{rect}(f_p/f_{ph}) = \begin{cases} 1, & f_p \subset V_f, \\ 0, & f_p \not\subset V_f; \end{cases} \qquad (6.13)$$

and $H_r(f_p) = \exp[j2\pi(f_{po} + f_p)|r_a|]$ is the transfer function of the compensation step of the target radial displacement.

The process of image reconstruction from a hologram described by Eq. (6.11) can be represented as

$$\hat{g}(\vec{r}_{no}) = F^{-1}\{S(\vec{f}_p) H_f(\vec{f}_p)\} = \int_{V_f} S(\vec{f}_p) H_f(\vec{f}_p) \exp(j2\pi \vec{f}_p \vec{r}_{no}) \, d\vec{f}_p$$

$$= g(\vec{r}_{no}) * h_o(r_{no}), \qquad (6.14)$$

where $h_o(\vec{r}_{no}) = F^{-1}\{H_o(\vec{f}_p)\}$ is a perfect impulse response which only describes the image noise due to the finite diffraction limit, or to the limited size of the aperture function $H_o(\vec{f}_p)$.

Thus, the processing of an echo signal during the imaging includes two stages (Fig. 6.3). The signal preprocessing is aimed at synthesising a Fourier hologram, whose size and shape are determined by the transmitter pulse parameters and the target aspect variation. The structure and composition of processing operations 1–5 are conventional radar operations and can be varied with the type of transmitter

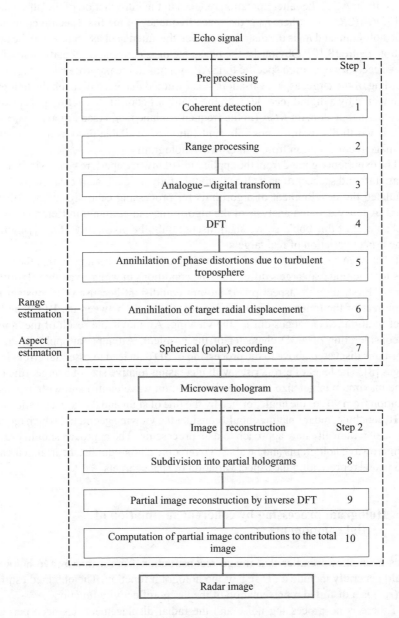

Figure 6.3 *The sequence of operations in radar data processing during imaging*

signal, the processing techniques used, and the tracking conditions. For example, a monochromatic pulse does not require operations 2 and 4. When a signal with a LFM is subjected to correlated processing, operations 1 and 2 coincide, and operation 4 becomes unnecessary. The compensation for the radial displacement of a satellite during hologram recording in field conditions is a fairly complex problem [8,10]. In an AEC, the latter operation reduces to the introduction of the phase factor $\exp\lfloor j2\pi f_{p1}(2R_{o}/c)\rfloor$, where f_{p1} is the space frequency of the first spectral component of the hologram and R_0 is the distance between the antenna phase centre and the target rotation centre [8,10]. Obviously, the phase factor is constant for a particular AEC.

A necessary operation specific to ISAR systems at the preprocessing stage is the recording of the target aspect variation. It is assumed that each pixel on the hologram is compared by a digital recorder with the family of coordinates defining its position in the frequency domain $f_x f_y f_z$ (in the frequency plane f_x–f_y) (see Fig. 6.2).

It is worth discussing a possible application of available processing algorithms for image reconstruction from a microwave hologram.

The experience gained from the application of inverse aperture synthesis for imaging aircraft and spacecraft as well as from the study of local radar characteristics has stimulated the development of algorithms for processing echo signals by coherent radars. A fairly detailed analysis of the algorithms can be found in Reference 8 and in Chapter 2 of this book, so we shall discuss only the possibility of applying them to the aspect variation of real targets.

It has been shown in Section 2.3.2 and in the References 9 and 10 that the conditions for tracking real targets differ from the conditions in which available algorithms operate. First, discrete aspect pixels are not equidistant because of a constant repetition rate of the transmitter pulses. Second, the angle between the RLOS and the target rotation axis changes during the viewing. An inevitable result of the latter is the consideration of a 3D character of the problem. Attempts at applying the 2D algorithms discussed above to the processing of 3D data lead to essential errors in the images [8]. The level of errors rises with increasing relative size of a target (the ratio of the maximum target size to the carrier radiation wavelength) and with increasing deviation from 90° of the angle formed by the line of sight and the target rotation axis.

To conclude, radar imaging should consider the viewing geometry, which requires the use of a radically new approach to data processing. The approach should provide 3D microwave holograms and be able to overcome a non-equidistant arrangement of echo pixels representing the aspect variation of space targets.

6.3 Hologram processing by coherent summation of partial components

It has been shown earlier that image reconstruction from a microwave hologram should generally include a 3D IFT of the hologram function. The obtained estimate of $\hat{g}(\vec{r}_{no})$ is a distorted representation of the target reflectivity function.

If there is no processing noise and the radial displacement has been perfectly compensated, an error may be due to a limited bandwidth of the transmitter pulse or a limited aspect variation. The resolving power of image-synthesising devices is then restricted only by the diffraction limit, and the image produced is known as a diffraction-limited image. Recording and processing noise additionally deteriorate image quality. So when designing algorithms and techniques for image processing, one should bear the following things in mind: (1) the dimensionality of an image is not to be higher than that of a microwave hologram and (2) the image resolution in any direction is to be inversely proportional to the hologram length. Hence, processing of 3D holograms can yield 1D, 2D and 3D images. An advantage of a 3D image is that it fully represents the information recorded on the hologram, but it is to be computer-processed and analysed. For visualisation, an image must be displayed on 2D media, such as paper or photosensitive films, or on computer screens. Moreover, the 'third' dimension of a hologram is sometimes insufficient to get a good resolution. Nonetheless, the neglect of a non-3D format of a hologram leads to serious image errors during its processing. Therefore, the problem of producing undistorted 2D images from 3D holograms seems quite important. We can suggest two ways of solving this problem.

One way is to obtain a 3D image and then intercept it with a plane of prescribed orientation. However, the computations with cumbersome 3D coordinates and data arrays of lower dimensionality require special processing algorithms and large computation resources.

A more simple and cost-effective approach is to compute directly the contributions of single 3D hologram components to a 2D image, if their dimensionality is not higher than that of the image. The computations become less complex and all highlighted components of a hologram can be processed simultaneously, provided that the number of processors is sufficient. The applicability of this technique can be easily extended to 2D holograms.

This method of image reconstruction can be termed coherent summation of partial components of a hologram. This method includes the following procedures:

Stage 1. A microwave hologram is subdivided into regions of limited size called partial holograms (PH). Since discrete pixels making up the hologram are formed by the interceptions of radial lines (corresponding to single echo signals) and cofocal spherical surfaces (corresponding to discrete values of the space frequency), PHs can be separated from the initial hologram in different ways.

The PH dimensionality is chosen from the initial hologram geometry and from considerations of processing convenience. In the case of a 2D hologram, the PHs may be one- or two-dimensional, while for a 3D hologram they may be, in addition, three-dimensional. Figures 6.4 and 6.5 depict 1D PHs with lines having points at their ends, which represent the initial and final pixels. The points on the surfaces of 2D and 3D PHs correspond to single pixels.

One-dimensional PHs are composed either of pixels coinciding with the radial rays which correspond to single pulses (radial PHs) or of pixels located on the cofocal spherical surfaces with f_{po} = const. (transverse PHs). Radial 2D PHs are made up of ensembles of 1D radial PHs and represent regions of planar conic (Fig. 6.5(b)) or more complex curved (Fig. 6.4(b)) surfaces. Transverse 2D PHs

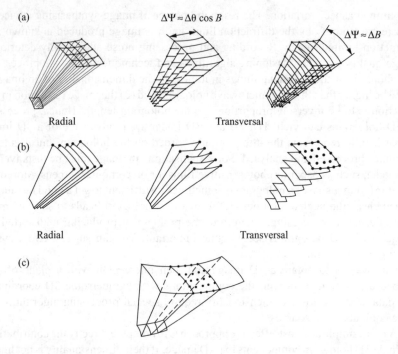

Figure 6.4 Subdivision of a 3D microwave hologram into partial holograms: (a) 1D partial (radial and transversal), (b) 2D partial (radial and transversal) and (c) 3D partial holograms

can be separated only from volume holograms. They are regions of spherical surfaces with f_{po} = const.

If the angular discretisation of a hologram is uniform, the maximum angle of 1D transverse, 2D and 3D PHs are chosen from the following considerations. When a spherical coordinate grid (or a polar grid for plane holograms) is replaced by a rectangular grid, the phase noise at the PH edges should not exceed $\pi/2$. This criterion leads to the following restrictions:

$$\Delta \psi \leq (\lambda/D)^{1/2}, \tag{6.15}$$

$$\Delta \psi \leq c/D\Delta f. \tag{6.16}$$

If the intersample spacing on a hologram varies slowly because of a non-uniform rotation of a target, the choice of the PH angle should meet the condition:

$$\delta \nu' \leq f \arccos(1 - \lambda/4r_n \cos \beta), \tag{6.17}$$

where $\delta \nu'$ is the difference between the maximum (or minimum) discretisation step and its average value. Condition (6.17) is based on the limited phase noise due to the non-equidistant arrangement of the hologram samples.

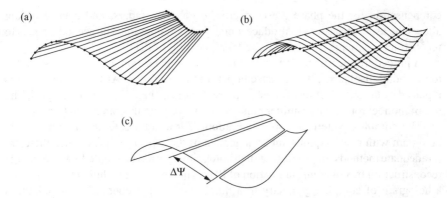

Figure 6.5 Subdivision of a 3D surface hologram into partial holograms: (a) radial, (b) 1D partial transversal and (c) 2D partial

When choosing the PH angle, one should always follow the more rigid of the above criteria. The restriction on the PH size is introduced in order to keep the deviation of the hologram samples from the rectangular grid nodes within a prescribed limit. The PH angles can be easily calculated analytically at a constant or slightly varying value of one of the angles of the spherical coordinates describing the PHs (Fig. 6.4(a)). In that case the PH boundaries will be close to the coordinate surfaces. If both angles θ and B change markedly (Fig. 6.5), the angular step $\Delta\psi$ should be found in the plane tangent to the PH.

Stage 2. Every PH should be subjected to a DFT providing a radar image with the same dimensionality as that of the PH, while the resolution is determined by its size.

Stage 3. The contributions of partial images to the integral image are computed. When the dimensionalities of a PH and a partial image are the same, the pixels of the latter are interpolated to those of the integral image. If the dimensionality of the integral image is higher, the major procedure for the computation is that of back projection [127].

Consider algorithms for the reconstruction of 2D images by processing narrow and wideband surface holograms (Fig. 6.5) produced by a three-axially stabilised ground radar. With such algorithms we shall try to justify the specific features of coherent summation of partial components: (1) the possibility of highlighting partial regions of various shapes on a PH and their independent processing and (2) the possibility to increase the resolution of the integral image as the individual contributions of the partial components are accumulated and the diffraction limit corresponding to the initial hologram size is achieved.

The above analysis allows the following conclusions to be drawn. The most general approach to radar imaging of a satellite by inverse aperture synthesis, no matter how it moves and what probing radiation is used, includes two stages of echo signal processing. The preprocessing involves some conventional operations, the compensation for the phase noise specific to coherent radars, and data recording allows the aspect variation to produce a microwave hologram. The second stage is to reconstruct the image by a special digital processing of PHs.

A procedure specific to preprocessing is the compensation for the phase shift due to the radial displacement of a space target. In the case of an AEC, this operation is replaced by the introduction of constant phase factors in the wideband echo signal. The use of monochromatic transmitter pulses does not require this operation (Chapter 5).

The complex pattern of aspect variation of low orbit satellites requires a 3D hologram with a non-equidistant arrangement of the aspect samples. Since there are no adequate methods for processing such holograms, we have designed a way of image reconstruction by coherent summation of PHs. This reduces the digital processing of a hologram of complex geometry to a number of simple operations. A hologram is subdivided into PHs, from which partial images are reconstructed using a fast Fourier transform (FFT). The contributions of the partial images to the integral image are computed.

6.4 Processing algorithms for holograms of complex geometry

We should first change Eq. (2.38) generally relating the hologram and image functions to the Cartesian coordinates necessary for a DFT:

$$\vec{f}_p \vec{r}_{no} = f_x r_x + f_y r_y + f_z r_z, \tag{6.18}$$

$$\text{where } f_x = |\vec{f}_p| \sin\theta \cos B, \tag{6.19}$$

$$f_y = -|\vec{f}_p| \cos\theta \sin B, \tag{6.20}$$

$$f_z = |\vec{f}_p| \sin B; \tag{6.21}$$

$$r_x = |\vec{r}_{no}| \sin\nu \cos\beta, \tag{6.22}$$

$$r_y = -|\vec{r}_{no}| \cos\nu \cos\beta, \tag{6.23}$$

$$r_z = -|\vec{r}_{no}| \sin\beta. \tag{6.24}$$

The substitution of Eq. (6.18) into Eq. (2.38) reduces it to the conventional 3D Fourier transform. However, it is impossible to apply it directly to a microwave hologram recorded in spherical coordinates (Fig. 6.2(b)). The transition to pixels located at rectangular grid nodes is considered as an interpolation problem. Even a first-order interpolation for a 2D case would require large computational resources. Besides, any noise arising from the interpolation would lead to large errors in the reconstructed image.

The procedure of coherent summation of partial components will simplify this problem if we use the reverse order of computational operations: a number of DFT operations and the interpolation of their results (partial images) to the rectangular grid nodes of the integral image. Of special practical importance is the case when a PH and its partial image have a lower dimensionality than the integral image. This is due to a higher computation efficiency of the algorithms used. The interpolation then represents a transition from a rectangular grid of lower dimensionality to that of a higher dimensionality, a procedure known as back projection [127].

As previously mentioned, we shall focus on designing algorithms for producing 2D images by coherent summation of 1D PHs and individual initial hologram samples. The algorithm for coherent summation of 2D partial images will largely be discussed for a theoretical completeness of the treatment. The analysis will start with algorithms for processing 2D holograms recorded in an AEC and during the imaging of low orbit satellites by a SAR located in the orbit plane.

6.4.1 2D viewing geometry

Equation (6.14) will be transformed to polar coordinates by substituting Eq. (6.18) into it and using Eqs (6.19)–(6.24). Assuming $B = \beta = 0$ and denoting $|\vec{f}_p| = f_{po} + f_p$ and $|\vec{r}_{no}| = r$, we get:

$$\hat{g}(r, \nu) = \int_{\theta_i}^{\theta_f} \int_{f_{pl}}^{f_{pu}} S(f_{po} + f_p, \theta)|f_p| \exp[j2\pi(f_{po} + f_p)r\cos(\nu - \theta)] \, df_p \, d\theta, \quad (6.25)$$

where θ_i and θ_f are the initial and final values of the angle θ of the hologram (Figs 6.6 and 6.7), $f_{pl} = f_{po} - \Delta f_p/2$ and $f_{pu} = f_{po} - \Delta f_p/2$ are the lower and upper boundaries of the space frequency band along the hologram radius.

It is easier to start the analysis of processing algorithms with a simple case of narrowband microwave holograms. The limit of expression (6.25) at $\Delta f_p \to 0$ is

$$g(r) = f_{po} \int_{\theta_i}^{\theta_f} S(f_{po}, \theta) \exp[j2\pi f_{po} r \cos(\nu - \theta)] \, d\theta. \quad (6.26)$$

This expression coincides with the formula for the CCA for a narrowband signal [94]. When an image is reconstructed by this algorithm, circular convolution is performed for every sample of the polar coordinate r in the image space with respect to the parameter θ of the hologram function and the phase factor. The contribution of all hologram samples to every (r, ν) node of the image polar grid is computed. If the satellite aspect changes non-uniformly, the samples are arranged along the hologram circumference with a variable step, so a discrete circular convolution becomes impossible.

Let us single out a series of adjacent regions on a hologram, or PHs shown in Fig. 6.6(a), with an angle satisfying the condition of Eq. (6.15). The convolution step of Eq. (6.26) over the whole hologram angle can be represented as a sum of integrals, each taken over a limited angle step $\Delta\theta$:

$$\hat{g}(r, \nu) = f_{po} \sum_{m=1}^{M} S_m(f_{po}, \theta) \exp[j2\pi f_{po} r \cos(\nu - \theta)] \, d\theta, \quad (6.27)$$

where $S_m(f_{po}\theta)$ is the mth PH and M is the total number of such holograms.

132 Radar imaging and holography

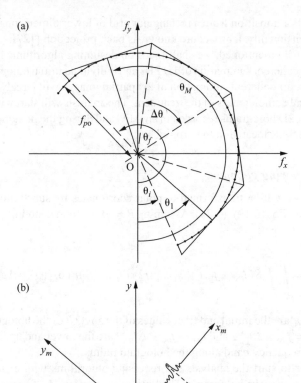

Figure 6.6 *Coherent summation of partial hologram. A 2D narrowband microwave hologram: (a) highlighting of partial holograms and (b) formation of an integral image*

We now introduce the Cartesian $x_m y_m$ coordinates (Fig. 6.6(b)) for each mth PH with the origin O coinciding with that of the rectangular x–y coordinates of the integral image. The x_m-axis is parallel to the tangent to the arc connecting the mth PH pixels at its centre. Since the microwave hologram in question is 2D, let us introduce the azimuthal coordinate $f_{p\theta} = f_{po}\theta$ to describe it in the frequency f_x–f_y plane (Fig. 6.6(a)), in addition to the radial polar coordinate f_p. With $x_m = r \sin \theta_m$ and $y_m = r \cos \theta_m$, the transformation of the phase factor under the integral of Eq. (6.27)

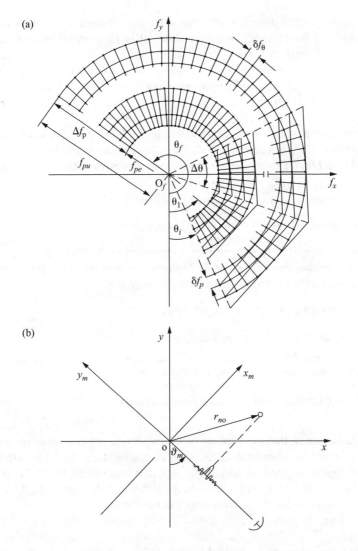

Figure 6.7   *Coherent summation of partial hologram. A 2D wideband microwave hologram: (a) highlighting of partial holograms, (b) formation of an integral image*

will give

$$\hat{g}(x,y) = \sum_{m=1}^{M} \int_{f_{\theta m}-\Delta f_\theta/2}^{f_{\theta m}+\Delta f_\theta/2} S(f_{po}, \theta) \exp[j 2\pi f_\theta x_m] \, df_\theta \, \Phi_m, \qquad (6.28)$$

where $df_\theta = f_{po} \, d\theta$ is the differential of the space frequency f_θ, while $f_{\theta m}$ is the space frequency corresponding to the mth PH centre.

Expression (6.28) describes the algorithm for coherent summation of partial images obtained from 1D transverse (azimuthal) PHs. Each partial image results from a Fourier transformation of the appropriate PH and is resolved along the azimuthal x_m-coordinate. The synthesis of a PH is made simultaneously with its summation with the radar image by moving the partial image along the y_m-coordinate (back projection), accompanied by the multiplication of all of its samples by a coherent processing phasor Φ_m.

The process of image summation by the algorithm of Eq. (6.28) will be discussed with reference to a point scatterer with the x_n, y_n coordinates in the x–y coordinate system (Fig. 6.6(b)). This scatterer will be assumed to possess an isotropic local radar target characteristic $g(\vec{r}_{no}) = \sigma_n^{1/2} \exp(j\varphi_n)$.

A narrowband microwave hologram is defined as

$$S(f_{po}, \theta) = \sigma_n^{1/2} \exp[-j2\pi f_{po}\hat{r}_n(\vartheta)]\exp(j\varphi_n). \tag{6.29}$$

The relative range of the point scatterer is expressed by the rectangular x_m, y_m coordinates. The expansion of $\hat{r}_n(\vartheta)$ into a Taylor series with respect to the centre ϑ_m of the mth partial angle step with the linear terms only gives

$$\hat{r}_n(\vartheta) = \hat{r}_n(\vartheta_m) + \dot{\hat{r}}(\vartheta_m)\vartheta - \dot{\hat{r}}_n(\vartheta_m)\vartheta_m, \tag{6.30}$$

where $\dot{\hat{r}}_n(\vartheta) = d\hat{r}_n(\vartheta)/d\vartheta$, $\dot{\hat{r}}_n(\vartheta_m) = \dot{\hat{r}}_n(\vartheta)|_{\vartheta=\vartheta_m}$.

By substituting Eq. (6.30) into Eq. (6.29) and denoting $\hat{r}_n(\vartheta) = y_m, \dot{\hat{r}}_n(\vartheta_m) = x_m$, we transform the expression for the mth PH to

$$S_m(f_{po}, \theta) = \sigma_n^{1/2} \exp\{-j2\pi f_{po}(y_{mn} - x_{mn}\vartheta_m)\}\exp(-j2\pi f_{po}x_{mn}\vartheta)\exp(j\varphi_n). \tag{6.31}$$

It is further assumed that the estimate of the target rotation rate obtained during the hologram recording contains no error: $\theta = \vartheta$. It should also be taken into account that a rectangular window of width $\Delta f_\theta = f_{po}\Delta\theta$ framing the PH (6.31) is shifted relative to the centre of the space frequency axis by its half width: $\Delta f_{po}\theta_m = \Delta f_\theta/2$. Then the expression for the partial image can be written as

$$\hat{g}(x_m) = \int_{f_{\theta m}-\Delta f_\theta/2}^{f_{\theta m}+\Delta f_\theta/2} S(f_{po}, \theta)\exp(j2\pi f_\theta x_m)\,df_\theta$$

$$= \sigma_n^{1/2}\Delta f_\theta\{\sin[\pi(x_m - x_{mn})\Delta f_\theta]/\pi(x_m - x_{mn})\}\exp[j\pi(x_m - x_{mn})\Delta f_\theta]$$
$$\times \Phi_{mn}\exp(j\varphi_n). \tag{6.32}$$

The integral image will be described as

$$\hat{g}(x, y) = \sigma_n^{1/2}\exp(j\varphi_n)\Delta f_\theta \sum_{m=1}^{M}\{\sin[\pi(x_m - x_{mn})\Delta f_\theta]/\pi(x_m - x_{mn})\Delta f_\theta\}$$
$$\times \exp[j\pi(x_m - x_{mn})\Delta f_\theta]\Phi_{mn}. \tag{6.33}$$

It is clear from Eq. (6.33) that the complex phase factors varying with the x_m, y_m coordinates and located at the integral image point corresponding to the position of the scatterer response have the maximum values equal to unity. The contribution of the PH to the integral image is defined by the product of the local radar target characteristic of the scatterer and the $\sin(x)/x$-type of function. Therefore, the PHs are summed equiphasically at the point $x_n = r_{no} \sin \vartheta_n, y_n = r_{no} \cos \vartheta_n$ and at other points, of the image, they are mutually neutralised.

The width of the major lobe of the scatterer in the partial image (a function of the $\sin(x)/x$-type) is determined by the PH length Δf_θ or by its angle $\Delta \theta$ (Fig. 6.6(a)). The limiting value of the response width in the partial image derived from Eq. (6.14) is expressed by the inequality $\delta x \geq 0.5(\lambda D)^{1/2}$. Since $D \gg \lambda$, the major lobe width is much greater than the transmitter pulse wavelength.

It follows from this treatment that the mth partial component of the integral image may be regarded as a 2D plane wave superimposed on the image plane. The wave front is normal to the y_m-axis and its period is equal to the half wavelength of the transmitter pulse. The initial wave phase (along the x_m-axis) is determined by the phasor $\exp[j\pi(x_m - x_{mn})\Delta f_\theta]$ in such a way that a positive half-wave always arrives at the scatterer's x_{mn}, y_{mn} position. The wave amplitude along the x_m-axis is described by a $\sin(x)/x$ function with a maximum at the point x_m. For this reason, the partial component has a 'comb' elongated by the back projection of the partial image parallel to the y_m-axis.

Note that the resolution of the integral image is defined by the scatterer wavelength rather than by the response width in the partial image. The reduction in the PH size from the maximum value prescribed by Eq. (6.15) to a single sample should not affect the result of summation in a PH. Therefore the synthesised aperture can be focused accurately over the whole image field. Keeping in mind

$$\lim_{\Delta f_\theta \to 0} \int_{f_{\theta m} + \Delta f_\theta/2}^{f_{\theta m} + \Delta f_\theta/2} S(f_{po}, \theta) \exp(j2\pi f_\theta x_m) \, df_\theta = f_{po} S(f_{po}, \theta) \, d\theta, \quad (6.34)$$

we obtain from Eq. (2.4) the algorithm for coherent summation of a PH made up of individual samples of the initial hologram:

$$\hat{g}(x, y) = f_{po} \sum_{m=1}^{M} S_m(f_{po}, \theta) \Phi_m. \quad (6.35)$$

The coherent summation algorithm for hologram samples essentially represents a particular case for 1D transverse (azimuthal) partial images described by Eq. (6.28). However, each has its own specificity.

The major advantage of the algorithm for hologram samples is the absence of phase errors due to either the PH approximation or the non-equidistant distribution of samples. As a consequence, this algorithm is applicable to the processing of microwave holograms with any known sample arrangement. On the other hand, the coherent summation algorithm for partial images does not require excessive computer resources because the exhaustive search of the raster pixels in the integral image

Table 6.1 The number of spectral components of a PH

Target size		Radial/azimuthal PH				
D (m)	$D_\lambda = D/\lambda$	$\mu = 0.02$	$\mu = 0.04$	$\mu = 0.06$	$\mu = 0.08$	$\mu = 0.1$
0.5	15	2/12	2/12	3/12	3/12	4/12
1.0	25	2/15	3/15	5/15	6/15	8/15
2.0	50	3/22	6/22	9/22	12/22	15/22
4.0	100	6/30	12/30	18/30	24/30	30/30
6.0	150	9/37	18/37	27/37	36/37	45/30
8.0	200	12/43	24/43	36/43	48/38	60/30
10.0	250	15/48	30/48	45/48	60/38	75/30
15.0	325	23/54	45/54	68/50	90/38	113/30

during the computation of the partial contribution is made for a group of PH samples rather than for every single hologram sample.

Figure 6.8–6.9 compares the computational complexity of the two algorithms as a function of the target size for a narrowband microwave hologram. The criterion for the degree of complexity is taken to be the algorithmic time of the programme realisation. The unit of measure of the algorithmic time is, in turn, taken to be 1 flop (floating point), that is, the time for one elementary operation of summation/multiplication of two operands with a floating point. So we have 1 Mflop $= 10^6$ flops. The estimations of the computational complexity and the programme realisation time have been made for a 2D image of 512 × 512 raster pixels in size and 2D microwave holograms with a 120° angle.

When going from a narrowband hologram to a wideband one, we can just suggest that the number of spectral components increases from 1 to L. As the size of a one-digit image and the hologram discretisation step are inversely proportional to each other, the minimal number of spectral components at a given pulse frequency bandwidth must be proportional to the target size. Table 6.1 presents the L values for various PHs as a function of the maximum target size. The computations have been made for 0.04 m carrier (centre) frequency of the transmitter pulse spectrum and the ratio of the image field size to the maximum target length $k = 1.5$. One can easily see that the number of azimuthal PH samples rises with the target size as long as the limiting PH angle obeys the inequality (6.15).

When a target is rather large and the relative frequency bandwidth is $\mu = \Delta f / f_0$ (the lower right-hand side of Table 6.1), the inequality (6.16) imposes a more rigid restriction on the PH size. Then both the PH size and its discretisation step decrease inversely with respect to the target size. Therefore, the number of PH azimuthal samples at a given transmitter pulse width Δf remains constant with increasing target size D.

We shall start the discussion of digital processing of 2D wideband holograms with the algorithm for coherent summation of 1D azimuthal partial images, which is the

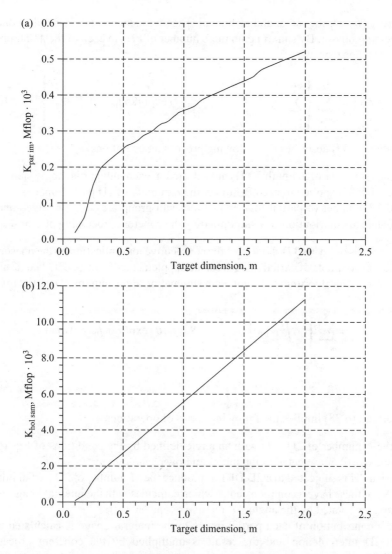

Figure 6.8 *The computational complexity of the coherent summation algorithms as a function of the target dimension for a narrowband microwave hologram: (a) transverse partial images, (b) hologram samples*

extension of a similar algorithm for narrowband microwave holograms. Let us relate Eq. (6.28) to the first, $1 = 1\text{th}, \ldots, L\text{th}$ spectral component:

$$\hat{g}(x,y) = \sum_{m=1}^{M} \int_{f_{\theta m}-\Delta f_\theta/2}^{f_{\theta m}+\Delta f_\theta/2} S_m(f_{pl},\theta) \exp(j2\pi f_{\theta l}x_m)\, df_{\theta l} \Phi_{ml}, \qquad (6.36)$$

where $f_{pl}, f_{\theta l}$ are the radial and azimuthal space frequencies and Φ_{ml} is the coherent processing phasor. By summing up the L number of PHs in each of the M number of partial angle steps, we get

$$\hat{g}(x,y) = \sum_{m=1}^{M} \left\{ \sum_{l=1}^{L} f_{pl} \int_{f_{\theta m}-\Delta f_\theta/2}^{f_{\theta m}+\Delta f_\theta/2} S_m(f_{pl},\theta) \exp(j2\pi f_{p\theta}x_m) \, df_\theta \, \Phi_{ml} \right\}. \quad (6.37)$$

Equation (6.37) describes the following processing operations:

- the L number of azimuthal PHs are selected in each mth partial angle step;
- the DFT is applied to each PH to get the L number of 1D partial images;
- the L number of partial images in every mth group are back projected and the obtained contributions are multiplied by the coherent processing phasor Φ_{ml}.

The analysis of Eq. (6.37) shows that the consecutive multiplication by the phasor Φ_{ml} of the contributions of partial images can be supplemented with a DFT. The result is a new processing algorithm – the coherent summation algorithm for 2D partial images:

$$\hat{g}(x,y) = \sum_{m=1}^{M} \left\{ \int_{-\Delta f_p/2}^{\Delta f_p/2} |f_p| \left[\int_{f_{\theta m}-\Delta f_\theta/2}^{f_{\theta m}+\Delta f_\theta/2} S_m(f_p,\theta) \exp(j2\pi f_{p\theta}x_m) \, df_\theta \right] \right.$$

$$\left. \times \exp(j2\pi f_p y_m) \, df_p \right\} \Phi_m. \quad (6.38)$$

Algorithm (6.38) implies the following series of operations:

- the M number of 2D PHs with an angle defined by the conditions of Eqs (6.15) and (6.16) are selected in the initial microwave hologram;
- each PH is subjected to a 2D DFT to produce the M number of 2D partial images. All of these have a common centre which coincides with the integral image centre and are rotated by the angle $\Delta\theta$ relative to one another;
- the contribution of each partial image to the integral image is calculated using a 2D interpolation and the result is multiplied by the coherent processing phasor.

The last operation generally requires large computer resources. So we shall further refer to the coherent summation algorithm for 2D partial images only to preserve a theoretical completeness.

The advantages of coherent summation of individual samples discussed above for narrowband holograms are fully valid for wideband holograms as well. Equations (6.37) and (6.34) yield

$$\hat{g}(x,y) = \sum_{m=1}^{M} \left[\sum_{l=1}^{L} f_{pl} S_m(f_{pl},\theta) \Phi_{ml} \, d\theta \right]. \quad (6.39)$$

Among the wideband processing algorithms, the one described by Eq. (6.39) is the most simple but it requires a large number of arithmetic operations to be made because the processing is made online. The computational efficiency of this algorithm can be raised by using a 1D DFT along the mth hologram beam:

$$\hat{g}(x,y) = \sum_{m=1}^{M} \left\{ \int_{-\Delta f_p/2}^{\Delta f_p/2} S(f_p,\theta)|f_p| \exp(j2\pi f_p y_m)\, df_p \Phi_m \right\} d\theta. \qquad (6.40)$$

In accordance with the accepted classification, expression (6.40) is the algorithm of coherent summation of 1D radial (range) partial images. Its implementation involves the following processing operations:

- the hologram samples making up the mth radial PH are multiplied by the linear frequency function and are then subjected to DFT;
- the resulting 1D range partial image is back projected and the result is multiplied by the coherent processing phasor.

The algorithm of Eq. (6.40) has much in common with the narrowband algorithm for partial images in Eq. (6.28) but it also has some specific features. One is that the 1D image module of a single point scatterer is described by the so-called kernel function of computerised tomography [57], rather than by a function of the $\sin(x)/x$ type. Depending on the chosen approximation of the linear frequency function in Eq. (6.40), the kernel function is its Fourier image and may be described analytically in various ways. It always has the form of an infinite periodic function with one major lobe and side lobes decreasing with amplitude. Another specificity of this algorithm is that the back projection operation is performed along the x_m-axis. Still another characteristic of the algorithm of Eq. (2.40) is that the PH samples are arranged equidistantly along radial straight lines, so no restriction is imposed on the maximum size of a PH.

The relative computational complexities of wideband processing algorithms are compared in Figs 6.10–6.11. It is seen that the number of arithmetic operations always increases with the relative frequency bandwidth of the transmitter pulse μ and the relative target size D_λ, whereas the computational complexity of 1D partial image algorithms changes differently with these parameters. At given values of μ and D_λ, more profitable is the algorithm for a PH with a larger number of samples. This is because the efficiency of a FFT increases with the number of lobes, as compared with an ordinary DFT. For example, at small values of μ and D_λ, it is more reasonable to use the algorithm for azimuthal partial images (Fig. 6.11). As the relative frequency bandwidth and the target size become larger, the number of samples in a radial PH exceeds, at a certain moment, that of an azimuthal PH (see Table 6.1). This happens because the restriction on the azimuthal PH size in Eq. (6.15) begins to dominate over that of Eq. (6.16), such that the use of the coherent summation algorithm for radial partial images becomes more profitable. In spite of its structural simplicity, the coherent summation algorithm for hologram samples has the greatest computational complexity (Fig. 6.10).

140 Radar imaging and holography

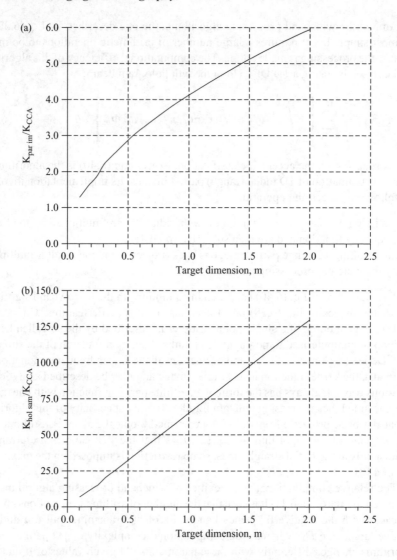

Figure 6.9 The relative computational complexity of coherent summation algorithms as a function of the target dimension for a narrowband microwave hologram: (a) transverse partial images/CCA, (b) hologram samples/CCA

It is clear that the time for a wideband hologram processing by the above algorithms, estimated from the product of the computational complexity and the time for an elementary multiplication/summation operation, is excessively long, so one should consider the possibility of separate, independent processing of PHs in order to considerably reduce this parameter.

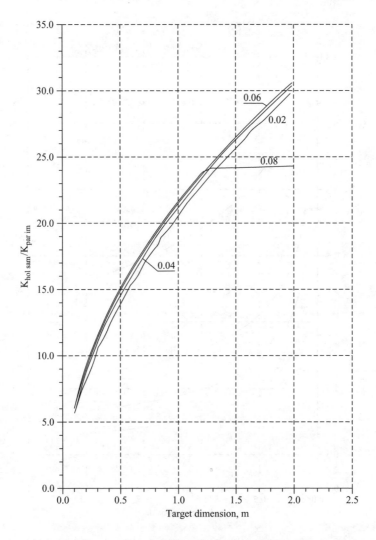

Figure 6.10 The relative computational complexity of coherent summation algorithms of hologram samples and transverse partial images versus the coefficient μ in the case of a wideband hologram

6.4.2 3D viewing geometry

We now express Eq. (6.14) in spherical coordinates and use Eqs (6.19)–(6.21) to get the relation

$$g(x,y,z) = \int\limits_{V_f} \iint S(f_p, \theta, B) \exp[j2\pi(f_{po} + f_p)(y\cos\theta\cos B$$

$$+ x\sin\theta\cos B + z\sin B)] \, df_f \, d\theta \, dB. \tag{6.41}$$

142 Radar imaging and holography

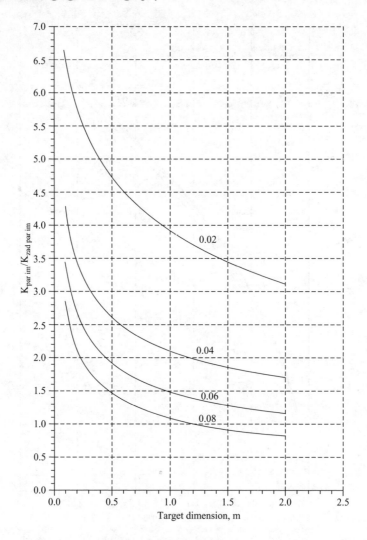

Figure 6.11 *The relative computational complexity of coherent summation algorithms for radial and transverse partial images versus the coefficient μ in the case of a wideband hologram*

To make the coherent summation of PHs more convenient, it is reasonable to separate the integration variables in Eq. (6.41). This task could be simplified if one of the variables remained constant through a synthesis step. For example, at $B = $ const., the image in the ($z = 0$) plane will be described as

$$g(x,y) = \int_{V_f} \iint S(f_p, \theta, B) \exp[j2\pi f_{pe}(y\cos\theta + x\sin\theta)] \, df_p \, d\theta, \qquad (6.42)$$

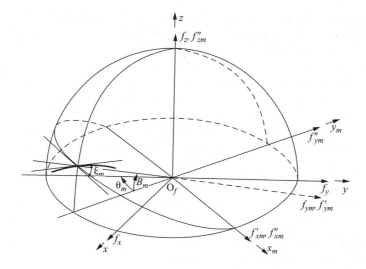

Figure 6.12   *The transformation of the partial coordinate frame in the processing of a 3D hologram by coherent summation of transverse partial images*

where $f_{pe} = (f_{po} + f_0 \cos B)$ is an 'equivalent' space frequency introduced just to reduce Eq. (6.42) to a conventional form.

Clearly, the algorithms to be derived from Eq. (6.42) may differ from those for 2D holograms only in the space frequency value. In reality, this may happen in viewing geostationary objects stabilised by rotation. In hologram recording of a low-orbit satellite, both angles describing its geometry change simultaneously. For this reason, the polar angle can be considered to be 'fixed' only at certain moments of time. This hologram geometry is best satisfied by the algorithms for coherent summation of individual hologram samples and 1D radial partial images. Expressions for such algorithms can be derived from Eq. (6.42) or directly from Eqs (6.35), (6.39) and (6.40) by substituting f_{pe} for $(f_{po} + f_p)$.

To design algorithms for transverse PHs, we need to introduce in the frequency domain the partial coordinates f_{xm}, f_{ym}, f_{zm} with the origin at the point O_f, which is also the origin of the f_x, f_y, f_z coordinates (Fig. 6.12). The f_{xm}–f_{ym} plane of the partial coordinates is tangential to the PH at a point with the angular coordinates θ_m, B_m (the f_{xm}- and f_{ym}-axes are not shown in Fig. 6.12).

Let us expand the polar angle B as a function of the azimuth θ into a Taylor series in the vicinity of θ_m:

$$B(\theta) = B_m + [dB(\theta)/d\theta]_{|\theta=\theta_m} \cdot (\theta - \theta_m) + \Delta_B, \tag{6.43}$$

where $B_m = B(\theta)$ at $\theta = \theta_m$ and Δ_B are the residual terms of the series. Obviously, in the ideal case of $\Delta_B = 0$, the PH lies totally in the f_{xm}–f_{ym} plane, and the difference between the PH and a straight line is only determined by the curvature of the sphere f_{po}. The non-zero nature of the polar angle derivatives with an order above the first one may generally lead to additional phase errors in the PH approximation by a straight line. However, a digital simulation of the aspect variation of a real target has shown that the phase error is negligible. Therefore, we shall assume the PH angle to be defined by the conditions of Eqs (6.15) and (6.16).

To describe the positions of PH samples in the f_{xm}–f_{ym} plane, we introduce the angle ψ and write down the partial Cartesian coordinates of the pixels as $f_{xm} = f_p \sin \psi$, $f_{ym} = -f_p \cos \psi$. An acceptable processing algorithm can be obtained if the f_{xm}–f_{ym} plane is superimposed with the f_x–f_y plane which corresponds to the x–y plane in the image space containing the integral image. The superposition operation will be made by two consecutive rotations of the partial coordinates (Fig. 6.12):

- the rotation by the angle $\xi_m = \mathrm{arctg}\lfloor (dB/d\theta)_{|\theta=\theta_m|}\rfloor$ round the f_{ym}-axis gives the polar $f'_{xm} f'_{ym} f'_{zm}$ coordinates, whose f'_{xm}-axis lies in the f_x–f_y plane;
- the rotation of the polar $f'_{xm} f'_{ym} f'_{zm}$ coordinates by the angle B_m round the f'_{xm}-axis gives the sought for polar $f''_{xm} f''_{ym} f''_{zm}$ coordinates.

These transformations of the polar coordinates result in the following expression for the scalar product at the mth partial angle step:

$$\lfloor \vec{f_p} \vec{r}_{no} \rfloor = -f_p b_m x'_m \sin \psi + f_{pe} y'_m \cos \psi \tag{6.44}$$

with

$$x'_m = x_m \cos \zeta_m + y_m \sin \zeta_m,$$
$$y'_m = -x_m \sin \zeta_m + y_m \cos \zeta_m,$$
$$b_m = (1 - \sin^2 \xi_m \cos^2 B_m)^{1/2},$$
$$f_{pe} = f_p \cos B_m.$$

In turn,

$$\sin \zeta_m = \sin \xi_m \sin B_m / b_m,$$
$$\cos \zeta_m = \cos \xi_m / b_m.$$

Thus, the variation of the polar angle B during hologram recording introduces two specific features in the coherent summation algorithm for transverse partial images. One is the necessity to make an additional rotation of the partial x_m, y_m coordinates by the ζ_m angle round the z_m-axis. The other is a change in the partial image scale along the x_m- and y_m-axes by a factor of b_m and $\cos B_m$, respectively.

Let us now derive an expression for the coherent summation algorithm for transverse partial images in the case of wideband pulses. This can be done by substituting Eq. (6.44) into Eq. (6.14) and reducing the result to the form:

$$\hat{g}(x,y) = \sum_{m=1}^{M} \left\{ \sum_{l=1}^{L} f_{ple} \int_{f_{\psi m}-\Delta f_\psi/2}^{f_{\psi m}+\Delta f_\psi/2} S(f_p, \theta, B) \exp(j2\pi f'_{\psi_o} x'_m) \, df_{\psi_o} \Phi_{ml\theta} \right\}, \tag{6.45}$$

Radar systems for rotating target imaging (tomographic approach) 145

where $f_{\psi o} = f_{po}\psi$ is the transverse space frequency; $f'_{\psi o} = f_{\psi o}b_m$; $f_{\text{ple}} = f_{pl}\cos B_m$ is the equivalent space frequency for the first spectral feature; and $\Phi_{ml\theta}$ is the coherent processing phasor.

The processing with the algorithm of Eq. (6.45) includes the following operations:

- the L number of transverse PHs (equal to the number of spectral features) are selected in every mth partial angle step;
- a DFT is performed with every PH with the space frequency f to produce the L number of 1D partial images;
- every partial image is back projected, and the contribution is multiplied by the phasor $\Phi_{ml\theta}$. The back projection is made along the y'_m-axis rotated by the ζ_m angle relative to the y_m-axis.

Equation (6.45) can be easily solved to give expressions for coherent summation algorithms for 2D partial and 1D transverse images of narrowband pulses, by analogy with the case discussed in Section 6.4.1.

As compared with the respective algorithms for 2D holograms, the computational complexity of coherent summation of individual hologram samples and 1D partial radial images increases only because of the necessity to compute the sine of the polar angle B. However, it does not increase more than by 2–3 per cent even in the most unfavourable case of a narrowband signal and a relatively small target. The complexity rises considerably when a 2D geometry is replaced by a 3D geometry of transverse partial images. This is due to both the polar angle variation during the viewing and the appearance of a variable in the hologram discretisation step. As a result, new operations come into play, but the increase in the computational complexity still lies within 10 per cent.

The above treatment allows the following conclusions to be made:

1. The algorithms for digital processing of microwave holograms designed in terms of the theory of coherent summation of partial components provide imaging in a wide range of viewing conditions, in particular, the probing geometry and the frequency bandwidth of transmitter radiation.
2. The wider applicability of digital processing by coherent summation of partial components implies a greater complexity of computations than that required by available techniques. However, one can choose the least time-consuming algorithm for particular values of the relative frequency bandwidth of the transmitter pulse and the size of the space target. A radical reduction in the processing time can be achieved by using separate processing of individual PHs.

Chapter 7
Imaging of targets moving in a straight line

When a target moves in a straight line normal to the radar line of sight, the inverse synthesis of a tracking aperture can be regarded in terms of Doppler information processing, in a way similar to the processing aimed at a high azimuthal resolution by a side-looking radar. Clearly, an inverse aperture can then be considered as a linear antenna array performing a periodic time discretisation of the radiation wave front. This is the so-called antenna approach, and its capabilities are discussed in Reference 139. The author analysed an equivalent array made up of $(2N+1)$ records of target movement across a real ground antenna beam of sufficient width. It was shown that the azimuthal range resolution R_0 and the resolution along the φ direction could be defined as

$$\Delta = \frac{\lambda R_0}{2VT_\mathrm{r}(2N+1)\cos\varphi}, \tag{7.1}$$

where λ is the transmitter pulse wavelength, V is the target velocity, φ is the angle between the line directed to the target and normal to the synthesising aperture, and T_r is the repetition rate of transmitter pulses.

Inverse aperture synthesis for a linearly moving target can also be examined in terms of a holographic approach. This was first done by H. Rogers to study ionosphere [85], making use of D. Gabor's ideas of holography. Rogers described a method for hologram recording of microwaves reflected by ionospheric inhomogeneities. The principle of this method is as follows. When an ionospheric inhomogeneity moves, the resulting diffraction pattern on the earth surface also moves across the receiver aperture. A signal that has been sensed is recorded on a photofilm as a hologram. What is actually recorded is the wave front, and one can reconstruct the inhomogeneity image from the hologram. For these reasons, E. Leith considered Rogers' device to be truly holographic rather than quasi-holographic.

Holographic concepts were successfully introduced in radar imaging by W. E. Kock [71] who showed that echo signals from a linearly moving target, recorded by the receiver of a coherent continuous pulse radar, were structurally equivalent to 1D holograms. He pointed out a similarity among an airborne SAR, a ground coherent radar and a holographic system.

The holographic approach treats inverse aperture synthesis of signals from a linearly moving target as a particular case of hologram recording by the scanning technique (Chapter 3). Here we shall analyse the process of radar imaging in the range–cross range coordinates, using inverse synthesis under real target flight conditions, that is, imaging of partially coherent signals.

Radar images obtained in the range–cross range coordinates allow estimate of the target size and shape, as well as the reflectivity of its individual scatterers. Such images can be further used for target identification. The imaging should be performed by ISARs transmitting complex pulses [85,104].

Apart from the prescribed movement, an aerodynamic target makes accidental motions with unknown parameters induced by destabilising factors, such as the constant component of wind velocity, the operation of the internal control system, turbulent flows, elastic fuselage oscillations and vibrations due to the engine operation and the target aerodynamics. Some of these can be estimated in advance by comparing the synthesis time T_s and the correlation time of perturbing effects T_c and by calculating the phase noise they introduce in the echo signal.

Among the above factors responsible for phase fluctuations of an echo signal $\psi(\varphi)$, of special importance are turbulent flows. This is because the constant wind velocity factor can be eliminated during the compensation for the radial displacement of a target. The second factor becomes important when a target is manoeuvring. For a typical synthesis time ($T_s \sim 1$ s), the value of T_c is smaller than that of T_s. The effect of the fourth factor can be avoided by choosing the wavelength λ such that the condition $\lambda/2 \gg \varepsilon$ (where ε is the maximum displacement due to fuselage oscillations) is fulfilled [17].

An echo signal from this kind of target is partially coherent. In the case of direct aperture synthesis, the effect of turbulent flows on the carrier pathway is accounted for by introducing a phase correction in the echo signal, which is found from random radial velocity and acceleration measurements [136]. In inverse synthesis, it is very hard to correct phase fluctuations $\psi(\varphi)$ of an echo signal. Below, we shall try to define the imaging conditions, primarily along the cross range coordinate, for a partially coherent signal [89]. The numerical simulation we have made shows that the destabilising factors of interest do not affect the range resolution.

7.1 The effect of partial signal coherence on the cross range resolution

Assuming that $f(x)$ is the distribution of the complex scattering amplitude (the target reflectivity) along the cross range x-coordinate, φ is an angle characterising the aspect variation, and $z(\varphi)$ is an echo signal, we have

$$z(\varphi) = \int f(x) \exp\left(-j\frac{4\pi}{\lambda}\varphi x\right) dx. \tag{7.2}$$

After the reconstruction of the radar image, which reduces to a Fourier transform of the echo signal (7.2) with the weight function $w(\varphi)$ and intensity, we obtain

$$|v(s)|^2 = \iint f(x_1)f^*(x_2)U(s-x_1, s-x_2)\,dx_1\,dx_2 + \eta(x), \tag{7.3}$$

$$U(s_1, s_2) = \iint w(\varphi_1)w^*(\varphi_2)$$
$$\times \exp\left[j\psi(\varphi_1) - \psi(\varphi_2) + \frac{j4\pi}{\lambda}(s_1\varphi_1 - s_2\varphi_2)\right] d\varphi_1 d\varphi_2, \tag{7.4}$$

where s is the cross range coordinate in the image plane, the sign * indicates complex conjugation, $\eta(x)$ is complex noise on the image, and $U(s_1, s_2)$ is the cross correlation function of the hologram.

The statistical characteristics $|v(s)|^2$ and $U(s_1, s_2)$ will be analysed on the assumption that $f(x)$ is a sum of the δ-functions of point scatterers and $\psi(\varphi)$ is defined by the normal distribution law. Consider the average $U(s_1, s_2)$ value over the phase fluctuations $\psi(\varphi)$, taking them to be Gaussian. With the formula for the characteristic function and the expansion of $\rho(\varphi_1 - \varphi_2)$ into a Taylor series at $\sigma^2 \gg 1$, we get

$$\langle \exp\{j|\psi(\varphi_1) - \psi(\varphi_2)|\}\rangle = \exp\{-\sigma^2[1 - \rho(\varphi_1 - \varphi_2)]\}$$
$$\cong \exp\{-\sigma^2/2\Delta^2(\varphi_1 - \varphi_2)^2\},$$

where σ^2 is the phase noise dispersion, $\rho(\varphi_1 - \varphi_2)$ is the correlation factor and Δ^2 is a quantity inverse to the second derivative of the correlation factor at zero, which describes the angle correlation step of the target aspect variation.

Assuming $w(\varphi) = \exp[-\varphi^2/(2\theta^2)]$, where θ describes the angle step of the synthesis, we find

$$\langle U(s_1, s_2)\rangle = \frac{\lambda^2 C}{64\pi\sqrt{(d_s^2 + d_c^2)^2 - d_c^4}}$$
$$\times \exp\left\{-\frac{d_s^2 + d_c^2}{2[(d_s^2 + d_c^2)^2 - d_c^2]} \cdot \left[(s_1 - s_2)^2 + 2\frac{d_s^2}{d_s^2 + d_c^2}s_1 s_2\right]\right\}, \tag{7.5}$$

where $C = \exp(-4\pi^2)$; $d_s = \lambda/(2\theta)$ is a resolution step corresponding to the synthesis time T_s (or the aspect variation $\theta = VT_s \sin\alpha/r_o$), V is the linear target velocity, α is the angle between the antenna pattern axis and the vector V, $d_c = \lambda\sigma/(2\Delta)$ is a space correlation step of target path instabilities, and r_o is the target range at the moment of time $T_s/2$.

The average intensity of a point target image (the impulse response of the system), derived for a partially coherent echo signal, is

$$\langle|v(s)|^2\rangle = \langle U(s-x_1, s-x_2)\rangle = C_1|A|^2 \exp\left(-\frac{(s-x)^2}{d_s^2 + 2d_c^2}\right), \tag{7.6}$$

where C_1 is the same factor of the exponent as in Eq. (7.5), A is the signal amplitude, and x is the scatterer coordinate.

For a target composed of a multiplicity of scatterers, each scatterer will be represented by a peak in the image described by Eq. (7.6). Its image position of each scatterer along the s-coordinate is its real position along the x-coordinate in the target plane. Moreover, every pair of scatterers will be represented in the image function by an interference term

$$\langle U(s-x_1, s-x_2)\rangle = C_1 \operatorname{Re} A_1 A_2^* \exp\left\{-\frac{[s-(x_1+x_2)/2]^2}{d_s^2 + 2d_c^2} - \frac{(x_1+x_2)^2}{4d_s^2}\right\}, \tag{7.7}$$

The additional term in Eq. (7.7) defines the peak located half way between the images of the respective scatterers; it has the same width as the peak for any other scatterer and is described by the ratio of the interscatterer distance to the resolution step value at zero phase noise. If this ratio is large, the interference term due to the superposition of side lobes in individual pixel images is negligible as compared with the average image intensity.

Under the conditions of partial signal coherence, the real resolution can be found from the 0.5 level of the maximum intensity $\langle |\nu(s)|^2\rangle$:

$$d_s' = 2s_{\langle|\nu(s)|^2\rangle=0.5} = C_2\sqrt{d_s^2 + 2d_c^2} = C_2 d_s\sqrt{1 + 2(\sigma T_s/\Delta)^2}, \tag{7.8}$$

where C_2 is a constant defined by the function $w(\varphi)$ in exponential and uniform approximations of $C_2 \cong 1.66$ and $C_2 = 1$, respectively. Obviously, if d_s decreases by the value Δ_s, the real resolution d_s' will improve only by Δ_s' (Fig. 7.1(a)):

$$\Delta_s' = C_2\sqrt{d_s^2 + 2d_c^2} - C_2\sqrt{(d_s - \Delta_s)^2 + 2d_c^2} \tag{7.9}$$

and with increasing T_s the gain in the real resolution will become still smaller.

Equation (7.9) can be reduced to

$$ad_s^2 + bd_s + c = 0, \quad \text{where } a = 4(p^2 - \Delta_s^2); \quad b = 4\Delta_s(\Delta_s^2 - p^2);$$
$$c = p^2(2\Delta_s^2 + 4d_c^2) - p^4 - \Delta_s^4; \quad p = \Delta_s'/C_2.$$

We can now calculate d_s and T_s values that may be considered most suitable for the synthesis at given Δ_s and Δ_s':

$$d_{s\,\mathrm{opt}} = \Delta_s/2 + \sqrt{\Delta_s^2/4 - c/a}, \tag{7.10}$$

$$T_{s\,\mathrm{opt}} = \lambda r_0/2V \sin\alpha d_{s\,\mathrm{opt}}. \tag{7.11}$$

At the values of $\lambda = 0.1$ m, $r_0 = 50$ km, $V = 600$ m/s, $\alpha = 90°$, $\Delta_s = 0.1$ m, $\Delta_s' = 0.05$ m and $C_2 = 1$, we find $T_{s\,\mathrm{opt}} = 1.83$ s for $T_c = 1.5$ s and $T_c = 3$ s, respectively ($d_c = 6.98$ m and $d_c = 3.49$ m).

Formula (7.11) defines the synthesis time of a partially coherent signal, which is optimal in the sense that it will require greater computer resources but will not essentially improve image quality determined by the real resolution d_s' or the Δ_s'/Δ_s

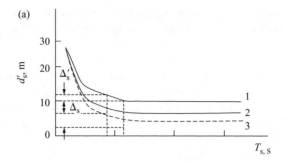

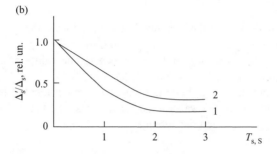

Figure 7.1 Characteristics of an imaging device in the case of partially coherent echo signals: (a) potential resolving power at $C_2 = 1$, (b) performance criterion ($1 - d_c = 6.98$ m, $2 - d_c = 3.49$ m and $3 - d_c = 0$)

ratio (Fig. 7.1(b)). This ratio quantitatively describes the gain in the angular radar resolution owing to the synthesis of partially coherent signals, as compared with that for perfect viewing conditions ($d_c \to 0$).

In the next section, we shall estimate the synthesis conditions by numerical simulation. The key factor in the imaging model to be described is target path fluctuations.

7.2 Modelling of path instabilities of an aerodynamic target

Path instabilities will be considered as random range displacements of a target (model I) or as independent fluctuations of the target velocity along the x'- and y'-axes (model II). The appropriate random processes will be expressed by recurrent difference equations [26]

$$Y_i[n] = \sum_{l=0}^{L} a_l X[n-l] + \sum_{k=0}^{K} b_k Y_i[n-k], \tag{7.12}$$

where the coefficients $a_0, a_1, \ldots, a_l$ and $b_1, b_2, \ldots, b_k$, as well as L and K vary with the cross correlation function; the subscript i denotes the number of the model for a random deviation of the target motion parameter. The coefficients a_l and b_k in Eq. (7.12) that are necessary for obtaining the values of $Y_i[n]$ with a prescribed coefficient $\rho(\tau)$ were presented in the work [26].

In model I of path instabilities, the current range $r_T[n]$ to a target is described as a sum of the predetermined range variation $r[n]$ and the random component $Y_l[n]$, which is the mean square deviation of the range. The quantities σ_p and T_c are: $\sigma_p = 0.04$ or 0.05 m, $T_c = 1.5$ and 3 s. The values of σ_p and T_c were found heuristically from a preliminary simulation.

In model II, the vector modulus of the real target velocity is

$$V_r[n] = \sqrt{(V + Y_{2x'}[n])^2 + Y_{2y'}^2[n]}, \qquad (7.13)$$

where $Y_{2x'}[n]$ and $Y_{2y'}[n]$ are the current values of random velocity deviations along the x'- and y'-axes, respectively; for comparison, the mean square deviation of the velocity is $\sigma_{x',y'} = 0.1$ or 0.2 m/s at $T_c = 1.5$ or 3 s. The values of $\sigma_{x',y'}$ and T_c are presented here courtesy of A. Bogdanov, O. Vasiliev, A. Savelyev and M. Chernykh who measured them in real flight conditions. Their experimental data on coherent radar signals in the centimetre wave range are also described in Reference 28.

The current angle between the antenna pattern axis and the vector $V_r[n]$, in this model, is

$$\alpha[n] = \alpha + \text{arctg}(Y_{2y'}[n]/(V + Y_{2x'}[n])). \qquad (7.14)$$

With Eqs (7.13) and (7.14) combined with the viewing conditions of model II, we have computed the real current range $r_T[n]$ to the target.

7.3 Modelling of radar imaging for partially coherent signals

To make the next step in the modelling of a radar image, we suggest that the predetermined path component of a point target is normal to the antenna pattern axis, that is, $\alpha = 90°$, the transmitter pulses have a spectral width $\Delta f_c = 75$ MHz, and their other parameters are chosen with the account of well-known restrictions for the removal of image inhomogeneities [104].

The range image of a target was formed by coherent correlation processing of every echo signal. For every pixel on the range image, the nth ($n = 1, \ldots, 256$) value of a complex echo signal was recorded to form a microwave hologram [138]. The reference function was formed ignoring the errors in the estimated parameters of target motion. The reconstructed image $|v(r,s)|^2$ was 2D in the r- and s-coordinates (range and cross range). The simulation showed that the phase noise due to path instabilities did not affect the range image of a target. Therefore, we shall further treat only its cross range section along the range axis.

A visual analysis of impulse responses during the imaging of partially coherent echo signals ($T_c = 3$ s, $T_s = 1.5$ s) indicates that phase fluctuations largely produce the following types of noise (Fig. 7.2). First, there is a shift of the impulse response along the s-axis in the image field (Fig. 7.2(a)). Second, the peak of the

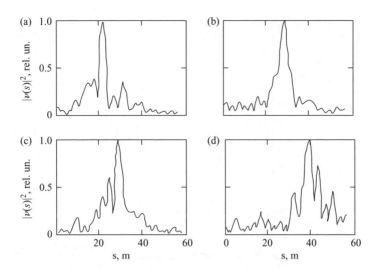

Figure 7.2 *Typical errors in the impulse response of an imaging device along the s-axis: (a) response shift, (b) response broadening, (c) increased amplitude of the response side lobes and (d) combined effect of the above factors*

major impulse response becomes broader (Fig. 7.2(b)). Third, the side lobes of the impulse response become larger to form some additional features commensurable in their intensities with the major peak (Fig. 7.2(c)).

Combinations of the three effects on the final image are also possible (Fig. 7.2(d)). It is worth noting that the first effect can be eliminated during the image processing by relating the window centre to the nth pixel with maximum intensity.

The presence of distorting effects necessitates finding ways to measure a real resolution step. A conventional way of estimating resolution is by measuring the impulse response of the processing device at the level 0.5 of the maximum intensity $|v(s)|^2$. In that case, analysis is made of all the images along the s-axis, independent of phase noise.

Another way of measuring a resolution step is that all additional features on a point target image at the 0.5 level are considered to be side lobes, irrespective of their intensity, and can be removed in advance.

Figures 7.3 and 7.4 present the estimates of an average resolution step d'_s for models I and II of path instabilities, respectively. The average value was calculated from 100 records of path instability of a point target for every discrete time moment T_s ($T_s = 0.1, \ldots, 2.9$ s). The estimation of a resolution step within model I fails to predict the degree of partial coherence effect on the radar image, since we know nothing about a perfect image *a priori*. The analysis of Fig. 7.3 has shown that the resolution step error is fairly large at $\sigma T_s/\Delta \geq 1$, where $\sigma = 2\pi\sigma_p/\lambda$. It is the appearance of false features above the 0.5 level with increasing synthesis time that leads to an overestimation of the resolution step computed from the impulse response

154 Radar imaging and holography

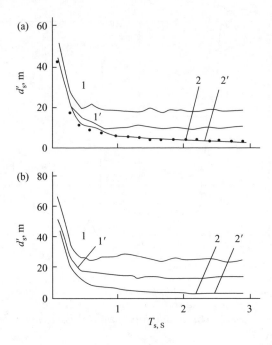

Figure 7.3 The resolving power of an imaging device in the presence of range instabilities versus the synthesis time T_s and the method of resolution step measurement: (a) $-\sigma_p = 0.04$ m; 1 and 1' (2 and 2') – first (second) way of resolution step measurement; 1 and 2 – $T_c = 1.5$ s, 1' and 2' – $T_c = 3$ s; (b) $-\sigma_p = 0.05$ m, 1 and 1' (2 and 2') – first (second) way of resolution step measurement; 1 and 2 – $T_c = 1.5$ s, 1' and 2' – $T_c = 3$ s

width and, hence, to a larger error in the target size measurement. Such an error is inherent in this method of resolution evaluation.

In the model of velocity instabilities (model II), the $d'_s(T_s)$ curves in Fig. 7.3 show a reasonable agreement with the theoretical curves in Fig. 7.1(a). The curve behaviour in Fig. 7.4 differs from the calculated dependences and from the model computations shown in Fig. 7.3 in that the $d'_s(T_s)$ curve has a minimum. The latter is due to an error in the method of estimating a resolution step, although the calculated $d'_s(T_s)$ curve does not indicate the presence of extrema.

The simulation results (curve 1' in Fig. 7.4(a)) can be used to find the synthesis time intervals for a particular type of signal (or a particular imaging algorithm): I – totally coherent, II – partially coherent and III – incoherent. One can choose various imaging algorithms for available statistical characteristics of path instabilities and for a particular time T_s. For instance, it is reasonable to use incoherent processing algorithms at synthesis times for which a signal can be considered as incoherent [78]. For shorter intervals I and II, one should use coherent processing algorithms and evaluate their performance in terms of the criterion Δ'_s/Δ_s (Fig. 7.5).

Imaging of targets moving in a line 155

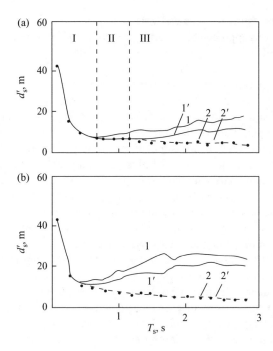

Figure 7.4 The resolving power of an imaging system in the presence of velocity instabilities versus the synthesis time T_s and the method of resolution step measurement: (a) $\sigma_{x'} = \sigma_{y'} = 0.01$ m/s (other details as in Fig. 7.3), (b) $\sigma_{x'} = \sigma_{y'} = 0.2$ m/s (other details as in Fig. 7.3)

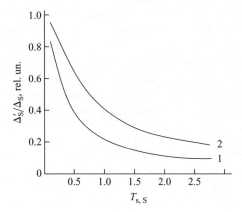

Figure 7.5 Evaluation of the performance of a processing device in the case of partially coherent signals versus the synthesis time T_s and the space step of path instability correlation d_c: $1 - d_c = 6.98$ m, $2 - d_c = 3.49$ m

The resolution estimate obtained by the second method is close to the theoretical value. However, this approach has a serious limitation because a real target possesses a large number of scatterers. The positions of respective intensity peaks on a radar image are unknown *a priori*, so the application of this technique may lead to a loss of information on adjacent scatterers on an image. This method proves to work well if one knows in advance that the target being viewed is a point object or that a range pixel corresponds to a single scatterer. In that case, the imaging device can be 'calibrated' by evaluating the phase noise effect on it.

The discrepancy between the simulation results presented in Figs 7.3 and 7.4 may be interpreted as follows. Model I of target path instabilities simulates random phase noise associated only with the displacement of range aperture pixels. Model II introduces greater phase errors in the echo signal, because the aperture is synthesised by non-equidistant pixels, which are additionally range-displaced. This model seems to better represent the real tracking conditions, since it accounts for random target yawing in addition to random range displacements.

The analytical expressions given earlier and the simulation results on partially coherent signals with zero compensation for the phase noise can provide the real resolving power of an imaging device. Today, there are no generally accepted criteria for evaluation of the performance of radar devices for imaging partially coherent signals. The results discussed in this chapter allow estimation of the device performance in the ideal case of $d_c \to 0$; on the other hand, they enable one to evaluate the efficiency of computer resources to be used in terms of the possible gain in the resolving power.

Track instabilities of real aerodynamic targets and other factors introducing phase noise give rise to numerous defects on an image. So the application of conventional ways of estimating the resolving power of imaging systems leads to errors. However, there is an optimal synthesis time interval which provides the best angular resolution with a minimal effect of phase fluctuations. Therefore, when phase noise cannot be avoided, which is usually the case in practice, it is reasonable to make use of a statistical database on fluctuations of motion parameters for various classes of targets and viewing conditions. The processing model we have suggested can be helpful in the evaluation of the optimal time of aperture synthesis in particular viewing conditions.

The viewing conditions also require a specific processing algorithm to be used, so radar-imaging devices should also be classified into coherent, partially coherent or incoherent. The simulation results presented in Fig. 7.4 do not question the validity of analytical relations (7.4), (7.5) and (7.7) but rather define their applicability, because a signal becomes incoherent when a fluctuating target is viewed for a long time.

Chapter 8
Phase errors and improvement of image quality

Possible sources of phase fluctuations of an echo signal, which negatively affect the aperture synthesis, are turbulent flows in the troposphere and ionosphere. Fluctuations of the refractive index due to tropospheric turbulence impose restrictions on the aperture centimetre wavelengths. Ionospheric turbulence affects far-decimetre wavelengths. Phase fluctuations decrease the resolving power of a synthetic aperture, leading to a lower image quality.

8.1 Phase errors due to tropospheric and ionospheric turbulence

8.1.1 The refractive index distribution in the troposphere

Fluctuations in the troposphere may arise from changes in the meteorological conditions and air whirls. As a result, there are non-uniform local distributions of temperature and humidity, leading to a non-uniform distribution of refractivity N:

$$N = (n-1) \times 10^6, \tag{8.1}$$

where n is the refractive index.

At the centimetre wavelengths, a static air volume has refractivity N defined by the Smith–Wentraub formula:

$$N = \frac{7.7P}{T} + \frac{3.73 \times 10^5 e}{T^2}, \tag{8.2}$$

where P is the total atmospheric pressure measured in millibars, T is temperature in Kelvin degrees and e is the specific water vapour pressure in millibars.

It follows from Eq. (8.2) that the value of N at centimetre and longer wavelengths strongly depends on the water vapour concentration, while its variation with the wavelength λ is insignificant. The latter fact is quite important because it makes it possible to obtain phase fluctuation spectra for various wavelengths in the microwave range, using an experimental spectrum measured at any wavelength. The major type of non-uniformity responsible for amplitude and phase fluctuations of an electromagnetic wave are so-called globules. These represent spherical or ellipsoidal structures, in which the refractive index differs, for some reason, from that in the environment. Generally, globules have arbitrary and irregular shapes. They arise from the local changes in the temperature, humidity or pressure accompanying turbulent phenomena in the troposphere. Since these causative factors behave differently at different points in space, the troposphere is generally non-uniform.

We shall first briefly describe the characteristics of a turbulent troposphere. The refractive index of the troposphere is generally the function $n(\vec{r}, t)$ of the radius vector $\vec{r}$ and time t, which can be written as

$$n(\vec{r}, t) = \langle n \rangle + \delta n(\vec{r}, t), \tag{8.3}$$

where $\langle n \rangle$ is an average value of the refractive index and $\delta n(\vec{r}, t)$ is its deviation from the average $\langle n \rangle$. Since the problem of interest is the fluctuation of the refractive index only, we shall further take $\langle n \rangle = 1$. The autocorrelation function of these fluctuations is

$$B_n(\vec{r}_1, \vec{r}_2, t_1, t_2) = \langle \delta n(\vec{r}_1, t_1) \delta n(\vec{r}_2, t_2) \rangle, \tag{8.4}$$

where $\vec{r}_1, \vec{r}_2$ are the radius vectors of the selected points.

For a steady-state turbulence, the autocorrelation function is independent of t (the steady state in time):

$$B_n(\vec{r}_1, \vec{r}_2) = \langle \delta n(\vec{r}_1, t) \delta n(\vec{r}_2, t) \rangle. \tag{8.5}$$

For a statistically non-uniform turbulence (the stationarity in space), the correlation function will not change if a pair of points $\vec{r}_1 u \vec{r}_2$ is displaced by the same distance and in the same direction simultaneously, that is, $B(\vec{r}_1, \vec{r}_2)$ varies only with $\vec{r}_1 - \vec{r}_2 = \vec{r}$. A spatially uniform distribution is called isotropic if $B_n(\vec{r})$ depends only on $r = |\vec{r}|$, that is, on the distance between the observation points but not on the direction.

However, even in the case of a uniform and isotropic random distribution of the refractive index, it appears to be quite difficult to choose an autocorrelation function for its fluctuations such that it could describe the real troposphere accurately. The only case when the fluctuation distribution can be described from theoretical considerations is a locally uniform isotropic turbulence. The general theory of this kind of turbulence was discussed in References 132 and 133. In real meteorological conditions, the distributions of wind velocity, pressure, humidity, temperature and the refractive index cannot be uniform or isotropic in large space regions. But in a relatively small region, whose size L_o is known as the outer-scale size of turbulence, the distributions may be taken to be both uniform and isotropic.

Theoretically, it is possible to describe fluctuations of the refractive index in terms of physical considerations of turbulence origin and development. The theory treats statistical fluctuations of velocity and related scalar quantities (such as temperature and the refractive index), induced by disturbances in horizontal air currents because of wind and by perturbations in laminar flow due to convection.

The physical mechanism of turbulence origin and development is as follows. When the translational wind velocity exceeds the critical Reynolds number, huge whirls (globules) arise and their size may exceed L_o. Such whirls are produced owing to the energy of translational flow movement, for example, to the wind power. This power is then given off to whirls of size L_o, and so on. Eventually, the energy is dissipated because of viscous friction in the smallest whirls of size l_o known as the inner-scale size of turbulence. In this way, huge whirls gradually split into smaller ones, and this process goes on until the power of rotational motion of the smallest whirls transforms to heat in overcoming the viscous force. For this reason, a region where huge whirls transform to small ones is called an inertia region. Within such a region, the instantaneous distribution of the refractive index $n(\vec{r})$ is an unsteady random function. However, the difference

$$n(\vec{r}_1) - n(\vec{r}_2)$$

is steady under the condition

$$|\vec{r}_2 - \vec{r}_1| < L_o.$$

In other words, $n(\vec{r})$ appears to be a random function with the first increments being steady. Random processes, like those discussed in the books [132,133], can be conveniently described by structure functions. The one for the refractive index distribution has the form:

$$D_n(r) = \langle [n(\vec{r}_1) - n(\vec{r}_2)]^2 \rangle. \tag{8.6}$$

The structure function is a fundamental characteristic of a random process with the first steady increments, replacing the concept of autocorrelation function. The latter just does not exist for random processes.

The quantity $D_n(r)$ describes the intensity of $n(r)$ fluctuations, whose periods are smaller or comparable with r. For a locally uniform and isotropic turbulence, it is defined as

$$D_n(r) = \langle [n(r_1 + r) - n(r_1)]^2 \rangle, \tag{8.7}$$

where r is an arbitrary increment of r_1.

Let us consider some statistical characteristics of the refractive index distribution in the troposphere. The detailed analysis made in References 132 and 133 has shown that the structure function of this parameter can be written as

$$D_n(r) = C_n^2 r^{2/3}, \qquad (l_o \ll r \ll L_o), \tag{8.8}$$

where C_n^2 is a structure constant of the refractive index. Equation (8.8) describes the so-called 2/3 law by Obukhov and Kolmogorov for the refractive index distribution. Numerous measurements made in the near-earth troposphere [132,133] showed a good agreement between the fluctuation characteristics of n and the 2/3 law. The value of l_o in the troposphere is found to be $\sim$1 mm. The quantity L_o is a function of direction and altitude. Therefore, one may assume that the horizontal extension of large whirls near the earth surface will have the same order of magnitude as the altitude, as far as the maximum altitudes lie in the range from 100 to 1000 m [110].

160 *Radar imaging and holography*

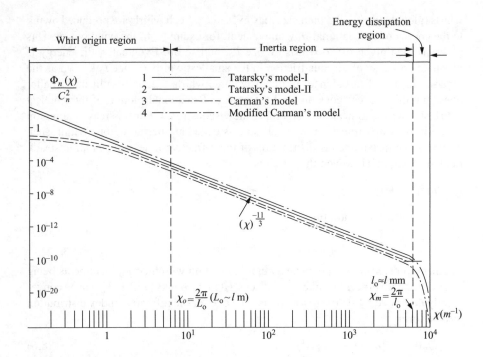

Figure 8.1   *The normalised refractive index spectrum $\Phi_n(\chi)/C_n^2$ as a function of the wave number χ in various models: 1 – Tatarsky's model-I, 2 – Tatarsky's model-II, 3 – Carman's model, 4 – modified Carman's model*

The refractive index spectrum obeying the 2/3 law is

$$\Phi_n(\vec{\chi}) = 0.033 C_n^2 \chi^{-11/3}, \qquad \text{at } (<\chi<\chi_m), \tag{8.9}$$

where $\chi_o \sim (2\pi/L_o)$, $\chi_m \sim (2\pi/l_o)$ and χ is the spatial wave number. It has been found experimentally that the $\Phi_n(\chi)$ spectrum has the form of $\chi^{-11/3}$ in an inertia region where the wave numbers are larger than χ_o. Figure 8.1 shows the normalised spectra for three regions: for the region of whirl origin ($\chi < (2\pi/L_o)$), for the inertia region ($(2\pi/L_o) \ll \chi \ll (2\pi/l_o)$) and for the dissipation region ($\chi \geq (2\pi/l_o)$).

It is seen that the spectral density $\Phi_n(\chi)$ in the region of $\chi \geq (2\pi/l_o)$ decreases much faster than might be expected from the $(\chi^{-11/3})$ formula. But in what way $\Phi_n(\chi)$ decreases in this region is still unclear theoretically. One usually deals with three kinds of spectra in the dissipation region. One obeys the $\chi^{-11/3}$ law, another drops abruptly at $\chi = \chi_m$, implying that $\Phi_n(\chi) = 0$ at $\chi = \chi_m$, and, finally, the spectrum changes on addition of the factor $\exp[-(\chi^2/\chi_m^2)]$.

The second case obeys Eq. (8.9) in practice. We have termed the respective model spectrum Tatarsky's model-I. It has been successfully employed in Reference 133 and some other studies. In Reference 132, V. Tatarsky used the following expression for the refractive index spectrum:

$$\Phi_n(\chi) = 0.033 C_n^2 \chi^{-11/3} \exp\left[-\frac{\chi^2}{\chi_m^2}\right] \qquad (8.10)$$

with $\chi_m/l_o = 5.92$ rather than 2π, as before. We have called the model for this case Tatarsky's model-II, which is fully valid in the inertia region but is approximate at $\chi > \chi_m$.

It follows from the analysis of the two models that they can adequately describe the statistical characteristics of the refractive index in the inertia region and are satisfactory for the dissipation region. In the region of $\chi < (2\pi/L_o)$, however, these models do not undergo any modification, that is, the dependence $\Phi_n(\chi)$ remains to be $\chi^{-11/3}$. On the other hand, it is known from References 132 and 133 that the spectral density curve $\Phi_n(\chi)$ at $\chi < (2\pi/L_o)$ is not universal and may change with the meteorological conditions. Therefore, the models of (8.9) and (8.10) are practically unable to evaluate the effects of this region on measurements. Besides, these models describe well only small-scale turbulence, which is quite clear from Fig. 8.1. In reality, however, most of the turbulence pulsation 'power' is accumulated in large whirls, at $\chi \leq (2\pi/L_o)$. In such regions, the uniformity and isotropic character of the random distribution of $n(\vec{r}, t)$ are also violated. Still, quantitative estimations can be made from interpolation formulae describing approximately the structure function behaviour at large L_o values, that is, in the range of small χ. One of these is Carman's function having the following spatial spectrum [133]:

$$\Phi_n(\vec{\chi}) = 0.063 \frac{\overline{\delta n_1^2} L_o^2}{(1 + \chi^2 L_o^2)^{11/6}} \qquad \text{at } \chi \ll \frac{2\pi}{L_o}, \qquad (8.11)$$

where $\overline{\delta n_1^2}$ is the dispersion of refractive index fluctuations.

The spectral model of (8.11) known as Carman's model works well for large-scale turbulence (Fig. 8.1). One can see from Eq. (8.11) that it does not include explicitly the constant C_n^2 related to the dispersion $\overline{\delta n_1^2}$ by the expression

$$C_n^2 = 1.9 \overline{\delta n_1^2} L_o^{-2/3}. \qquad (8.12)$$

Using Eq. (8.12), one can derive expressions for Tatarsky's models I and II:

$$\Phi_n(\vec{\chi}) = 0.063 \overline{\delta n_1^2} L_o^{-2/3} \chi^{-11/3} \qquad \text{at } \frac{2\pi}{L_o} \ll \chi \ll \frac{2\pi}{l_o}, \qquad (8.13)$$

$$\Phi_n(\vec{\chi}) = 0.063 \overline{\delta n_1^2} L_o^{-2/3} \chi^{-11/3} \exp\left[-\frac{\chi^2}{\chi_m^2}\right] \qquad \text{at } \frac{2\pi}{L_o} \ll \chi \ll \frac{2\pi}{l_o}. \qquad (8.14)$$

This representation is convenient when the refractive index fluctuations are given as $\overline{\delta n_1^2}$ rather than through C_n^2.

The next point to discuss is the applicability of the spectra described by Eqs (8.9), (8.10) and (8.11). When using this or that spectral model in problems of parameter fluctuations of an electromagnetic wave in a turbulent medium, one should bear in mind the following factors. First, the spectra are valid in the inertia region of a locally uniform and isotropic turbulence. Sometimes, the turbulence spectrum may strongly differ from the above models. Second, the spectrum at $\chi \leq \chi_o$ is, at best, an approximation, even though one may use Carman's spectra. At $\chi \geq \chi_m$, the model spectra are only good approximations. Note that the spectrum of the form (8.11) transforms to that of (8.9) at $\chi^2 L^2 \gg 1$. In addition to the three types of spectra, there is a spectrum of the form:

$$\Phi_n(\vec{\chi}) = \frac{\alpha \exp(-\chi^2/\chi_m^2)}{(1+\chi^2 L_o^2)^{11/6}},$$

$$\alpha = \frac{\overline{\delta n_1^2} L_o^3}{\pi^{3/2}} \frac{\Gamma(11/6)}{\Gamma(1/3)} C(\chi_m L_o),$$

$$C(\chi_m L_o) \approx \left[1 + \frac{\Gamma(11/6)}{\Gamma(1/3)} \frac{\Gamma(-1/3)}{\Gamma(3/2)} (\chi_m L_o)^{-2/3}\right]^{-1}. \tag{8.15}$$

At $\chi_m L_o \gg 1$, the correction term $C(\chi_m L_o) \approx 1$. Since $l_o \sim (1 \div 10)$ mm and $L_o \geq 1$ m, we have

$$\frac{\chi_m}{l_o} = 5.92, \qquad \chi_m = (5.92 \div 59.2),$$

$$\chi_m L_o \geq 5.92 \times 10^3.$$

Keeping in mind this fact and

$$\frac{\Gamma(11/6)}{\pi^{3/2}\Gamma(1/3)} \approx 0.06,$$

we get

$$\Phi_n(\vec{\chi}) = 0.06 \frac{\overline{\delta n_1^2} L}{[1+\chi^2 L_o^2]^{-11/6}} \exp\left(-\frac{\chi^2}{\chi_m^2}\right) \tag{8.16}$$

or

$$\Phi_n(\vec{\chi}) = 0.06 \frac{C_n^2 L_o^{11/3}}{[1+\chi^2 L_o^2]^{-11/6}} \exp\left(-\frac{\chi^2}{\chi_m^2}\right). \tag{8.17}$$

It would be reasonable to call a spectrum of the type (8.16) or (8.17) Carman's modified spectrum. If relation (8.12) is fulfilled, this spectrum will coincide with that described by Eqs (8.10) and (8.14) at large values of χ. But in the χ range, it coincides with the Carman spectrum shown in Fig. 8.1. The choice of a particular type of spectrum varies with the problem to be solved. Fluctuations of some electromagnetic wave parameters, such as phase and amplitude, are often sensitive to a certain turbulence spectrum, or to large- or small-scale whirls. Keeping this important fact in mind, one should analyse carefully the applicability of the chosen spectrum before using it.

The best way of verifying a model is to compare the results obtained with available experimental data. Although the models of (8.9) and (8.10) are rather approximate at $\chi < (2\pi/L_0)$, they still provide a good agreement with measurements (e.g. of phase fluctuations). Moreover, they can give the results in an analytical form. On the other hand, the models of (8.11) and (8.15) are more accurate for large whirls but they are unable to give clear analytical results. These circumstances have predetermined the applicability of the models of (8.8) and (8.10). In the study of phase fluctuations, both models yield similar analytical expressions.

It is of importance to discuss in some detail a vertical profile model of the structure constant. This constant describes the degree of refractive index non-uniformity, because it relates the quantities $D(r)$ and r (see Eq. (8.8)). The structure constant C_n^2 is related to the tropospheric parameters $\overline{\delta n_1^2}$ and r. For radiation propagation along an oblique path, the turbulence 'intensity' changes with the altitude, and the C_n^2 values will be different at different altitudes. The structure function of $n(r)$ will then be

$$D_n(r) = C_n^2(h) r^{2/3},$$

where $C_n^2(h)$ is a structure constant varying with altitude. To obtain quantitative results, one is first to find the $C_n^2(h)$ variation. The theoretical treatment of the problem of parameter fluctuations for a plane wave in a turbulent troposphere [132] included the following $C_n^2(h)$ models:

$$C_n^2 = C_{n0}^2 \exp\left(-\frac{h}{h_0}\right), \tag{8.18}$$

$$C_n^2 = C_{n0}^2 \frac{1}{1 + (h/h_0)^2}, \tag{8.19}$$

where C_{n0}^2 is the structure constant of the refractive index near the earth surface, h is the altitude of the point in question, and h_0 is a constant.

However, the question whether Eqs (8.18) and (8.19) can really describe the $C_n^2(h)$ function in the microwave frequency band remains unanswered. In order to find the exact form of this function, it is necessary to examine the microstructure of the refractive index distribution in the microwave range and to design a $C_n^2(h)$ model. This became possible only after the publication of the work [134], which reported measurements made in experimental flight conditions. The structure constant profile of the refractive index was measured along an oblique microwave path. The results of the $C_n(h)$ measurement were summarised in table 1 of Reference 134. Yet, it was impossible to plot the $C_n(h)$ function from these data, because they were to be statistically processed. This was accomplished by the authors of Reference 144.

Figures 8.2 and 8.3 show some $C_n^2(h)$ plots for different seasons (for April and November). The C_n^2 values in these plots represent records averaged over several runs of the squared structure constant measurement (the averaging was actually made over the time of day). The confidence limit was taken to be 0.98. Some of the C_n values presented in Reference 134 differ considerably from the average values and do not seem to be due to a statistical spread. To reveal such data, the authors used a

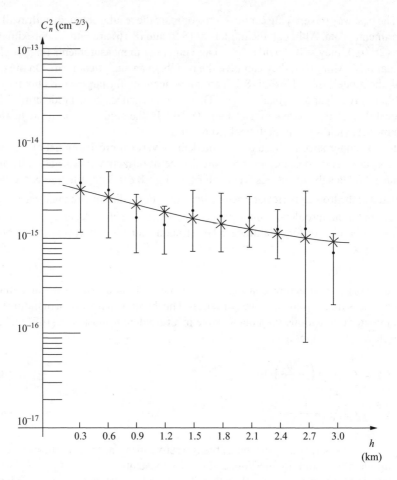

Figure 8.2 *The profile of the structure constant C_n^2 versus the altitude for April at the SAR wavelength of 3.12 cm*

criterion based on the assumption of a normal error distribution. The C_n records that differed from the average by more than a possible maximum of the statistical spread and were lying within the 0.98 error limit were eliminated from further analysis. The plots thus obtained were approximated by exponential functions, using the least square method. As a result, the following analytical dependencies were derived for the structure constant profile at the wavelength of 3.12 cm:

(a) the $C_n(h)$ model for April:

$$C_n^2(h) = C_{n0}^2 \exp\left(-\frac{h}{h_0}\right) \tag{8.20}$$

with $C_{n0}^2 = 3.69 \times 10^{-15}$ cm$^{-2/3}$ and $h_0 = 2.17 \times 10^5$ cm;

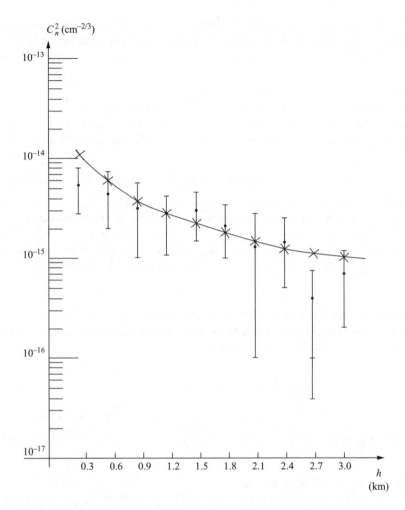

Figure 8.3 The profile of the structure constant C_n^2 versus the altitude for November at the SAR wavelength of 3.12 cm

(b) the $C_n(h)$ model for November:

$$C_n^2(h) = C_{n0}^2 \exp\left(-\frac{h}{h_0}\right), \tag{8.21}$$

with $C_{n0}^2 = 1.27 \times 10^{-14}$ cm$^{-2/3}$ and $h_0 = 8.89 \times 10^4$ cm.

We can see that the refractive index fluctuations decrease with altitude. The major contribution to the fluctuations is made by a tropospheric stratum 3 km thick above the earth. The contribution of the other 7 km thickness (the total thickness of the troposphere is taken to be 10 km) is five times smaller. It is known that the fluctuation of n increases with rising humidity. The most intense fluctuations are observed at the air–cloud interface and inside the clouds. This model, however, ignores these effects because of the lack of experimental data. But some data are available on the effect of humidity and clouds on the dispersion $\overline{\delta n^2}$ of the refractive index values. Therefore, the model of the vertical $\overline{\delta n^2}$ profile allows estimation, in a first approximation, of the cloud effect on phase fluctuations.

To conclude, it seems reasonable to extend the results on $\lambda = 3.12$ cm waves to other centimetre wavelengths, since the Smith–Wentraub formula (8.2) indicates only a slight dependence of n on the wavelength λ within the centimetre frequency band.

8.1.2 The distribution of electron density fluctuations in the ionosphere

In contrast to the troposphere, the ionosphere is characterised by electron density fluctuations. Let $\Delta Ne(\vec{r})$ denote fluctuations of the equilibrium electron density N_o, which is the average electron concentration. The variable ξ defined by the equality $\xi = \Delta Ne(\vec{r})$ represents a uniform random distribution with a zero average and a standard deviation σ_ξ. By definition, the autocorrelation function of this distribution is

$$B_\xi(\vec{r}_1 - \vec{r}_2) = \langle \xi(\vec{r}_1)\xi(\vec{r}_2) \rangle,$$

where the angular brackets stand for the averaging over an ensemble.

According to the Wiener–Khinchin theorem, the autocorrelation function and the spectrum create a Fourier transform pair:

$$\Phi_\xi(\vec{\chi}) = (2\pi)^{-3} \int\!\!\!\int\!\!\!\int_{-\infty}^{\infty} B_\xi(\vec{r}) e^{-j\vec{\chi}\vec{r}} d^3 r, \tag{8.22}$$

$$B_\xi(\vec{r}) = (2\pi)^{-3} \int\!\!\!\int\!\!\!\int_{-\infty}^{\infty} \Phi_\xi(\vec{\chi}) e^{-j\vec{\chi}\vec{K}} d^3 \chi, \tag{8.23}$$

where χ is the spatial wave number.

Experimental investigations have shown [141] that both the phase fluctuation spectra of a wave that has passed through a turbulent ionosphere and the amplitude fluctuation spectra have an asymptotic power dependence. Hence, the spectra of ionospheric whirls must also have a power dependence. Assuming the whirls to be isotropic within a space scale from 70 m to 7 km, a 3D whirl spectrum will have the form [141]:

$$\Phi_\xi(\vec{\chi}) \approx \chi^{-P}, \tag{8.24}$$

where P is the power index of the spectrum varying between $2 \leq P \leq 3$.

C. L. Rino and various co-workers [113–116] have suggested a spectrum of the electron density fluctuation:

$$\Phi_{\Delta Ne}(\vec{\chi}) = C_S \chi^{-(2\nu+1)} \quad \text{at } \chi_o < \chi < \chi_i, \tag{8.25}$$

where C_S is the turbulence parameter, χ_o is the outer-scale size of ionospheric whirls, χ_i is their inner-scale size, and $2\nu = P$ is the spectral power index. We have mentioned above that the power index varies between $2 < P < 3$, whereas the power index for the troposphere is $P = 8/3$ (Kolmogorov's spectrum).

The turbulence parameter is described as

$$C_S = 8(\pi)^{3/2}(\chi)^{P-2}\frac{\Gamma((P+1)/2)}{\Gamma((P-1)/2)}\langle \Delta Ne^2 \rangle \qquad (8.26)$$

where $\Gamma(\cdot)$ is the gamma-function and $\langle \Delta Ne^2 \rangle$ is the mean square value of the fluctuation component of the electron density. For a typical fluctuation distribution in the ionosphere, $C_S \sim 10^{21}$ (mKs). The quantity $\langle \Delta Ne^2 \rangle$ varies remarkably with the ionospheric conditions, so C_S fluctuates from 6.5×10^{19} (mKs) at $P = 2.9$ to 1.3×10^{23} (mKs) at $P = 1.5$ [22]. The ionosphere has a thickness of about 200 km. The maximum electron density lies in the Nm F2 stratum at an altitude between 250 and 350 km.

The outer-scale size of a turbulent whirl along the shortest distance (ionospheric whirls are anisotropic) is about 10 km. The respective value for a turbulent troposphere is about 1 km.

8.2 A model of phase errors in a turbulent troposphere

When discussing a SAR in Chapter 3, we pointed out that a turbulent non-uniform troposphere could be a source of spatial phase fluctuations. Let us consider a turbulence model with reference to a particular type of SAR – a radar with a focused aperture. Suppose a SAR is located along the carrier track (Fig. 8.4). For simplicity, we shall assume that there is only one point scatterer A across the swath width. This target is located at the point having an oblique range R and is scanned for the synthesis time L_s, i.e.

$$L_s \approx \beta H,$$

where β is the aperture pattern width.

The equiphase surface of an echo signal represents a sphere with the centre at the target location point. The track line is shown by the A_1–A_2 line, and the thickness of a turbulent tropospheric stratum is denoted as h_t. The structure function of the phase fluctuation for a spherical wave of the point target A is

$$D_\varphi(\rho) = \langle [\varphi(r+\rho) - \varphi(r)]^2 \rangle, \qquad (8.27)$$

where ρ is the distance between the points, at which the phase fluctuations are to be measured, for example, $\rho = d_e$. To find an analytical expression for $D(\rho)$, consider a 2D spectrum of wave phase fluctuations in a turbulent troposphere. Using a gradual perturbation approach, the authors of Reference 133 derived a simple formula relating the phase fluctuation parameters to the spectral density of the refractive index fluctuations $\Phi_n(\vec{\chi})$. The 2D spectral density $F_\varphi(\vec{\chi}, 0)$ and $\Phi_n(\vec{\chi})$ have the simplest relation, because the former is a 2D Fourier transform of the respective phase structure function in the plane $x = $ const. normal to the wave propagation direction. For a plane

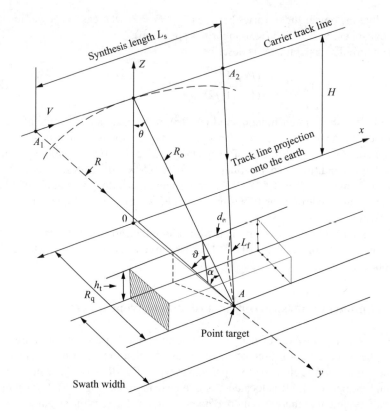

Figure 8.4 *A geometrical construction for a spaceborne SAR tracking a point object A through a turbulent atmospheric stratum of thickness h_t*

wave with the cross section $x = L$, we have

$$F_\varphi(\chi, 0) = \pi k^2 L \left[1 + \frac{k}{\chi^2 L} \sin \frac{\chi^2 L}{k} \right] \Phi_n(\vec{\chi}), \tag{8.28}$$

where L is the distance covered by the wave passing through a non-uniform turbulent medium.

Using Eq. (1.51) from Reference 133, we can now turn to the structure function of phase fluctuations in the plane $x = L$:

$$D(\rho) = 4\pi \int_0^\infty [1 - J_0(\chi\rho)] F_\varphi(\chi, 0) \chi \, d\chi, \tag{8.29}$$

where ρ is the distance between the points, at which the structure function is to be measured in the plane $x = L$. It follows from Eq. (8.28) that the 2D spectrum of $F_\varphi(\vec{\chi}, 0)$ is similar to the spectrum of the refractive index fluctuations $\Phi_n(\vec{\chi})$ multiplied by the filtering function (in square brackets). Therefore, the wave propagation through a turbulent medium is similar to the linear filter effect in circuit theory.

The filtering function of phase fluctuations is only slightly sensitive to the parameter variations. For example, at $\chi = 0$, $F_\varphi(\vec{\chi}, 0)$ is equal to $2\pi k^2 L$, changing smoothly with increasing χ as far as $\pi k^2 L$. Therefore, the filtering occurs relatively uniformly. The maximum product of the filtering function and $\Phi_n(\vec{\chi})$ for typical SARs is observed at small values of χ, or in large whirls. For this reason, phase fluctuations and phase correlation are most sensitive to the outer-scale size of turbulence, L_o.

With Eq. (8.29) and the turbulence models of (8.9) and (8.10), we can arrive at an expression for a uniform turbulence and a plane wave:

$$D_\varphi(\rho) = \alpha k_1^2 L^2 \rho^{5/3}, \tag{8.30}$$

where

$$\alpha = \begin{cases} 2.91, & \text{at } \rho \geq \sqrt{\lambda L}, \\ 1.46, & \text{at } l_o \ll \rho \ll \sqrt{\lambda L}, \end{cases}$$

and L is the electromagnetic wave path in a turbulent medium.

In order to examine the effect of phase errors on the recording of 1D holograms by a side-looking radar, it would be useful to try to extend the above result to the case of a non-uniform turbulence and a spherical wave [144].

From Tatarsky's non-uniform model-I, we have

$$D_\varphi(\rho) = 1.46 k^2 \rho^{5/3} \int_0^L C_n^2(h)\, dh, \quad (l_o \ll \rho \ll \sqrt{\lambda L}), \tag{8.31}$$

$$D_\varphi(\rho) = 2.91 k^2 \rho^{5/3} \int_0^L C_n^2(h)\, dh, \quad (\rho \gg \sqrt{\lambda L}). \tag{8.32}$$

The last two expressions show that phase fluctuations are equally affected by all whirls, irrespective of their distance to the observation point. Moreover, when ρ passes through the value $\sqrt{\lambda L}$, which is usually somewhere at the beginning of the path, the factor in front of $D_\varphi(\rho)$ increases 2-fold. Therefore, the experimental structure function $D_\varphi(\rho)$ must have a positive rise at $\rho = \sqrt{\lambda L}$.

It is interesting to follow how $D_\varphi(\rho)$ changes when a plane wave is replaced by a spherical one. The formula relating the mean square value of the phase difference fluctuation to the base 'ρ' for a spherical and plane wave [132] is

$$\left\langle (\varphi_1 - \varphi_2)_{\text{sp}}^2 \right\rangle = [D_\varphi(\rho)]_{\text{sp}} = \int_0^1 D_\varphi(\rho t)\, dt.$$

For the plane wave $D_\varphi(\rho) = \alpha k^2 C_n^2 L \rho^{5/3}$, we have

$$[D_\varphi(\rho)]_{\text{pl}} = A_0 \int_0^1 D(t\rho)^{5/3}\, dt = \frac{3}{8} A_0 \rho^{5/3},$$

where $A_0 = \alpha k^2 C_n^2 L$. Hence,

$$[D_\varphi(\rho)]_{sp} = \frac{3}{8}[D_\varphi(\rho)]_{pl}. \tag{8.33}$$

We can conclude that phase fluctuations for a spherical wave are not as large as for a plane wave and that the structure functions for the former differ from those of the latter only in numerical coefficients. For a medium with slowly changing characteristics, we have

$$[D_\varphi(\rho)]_{sp} = \frac{3}{8} 1.46 k^2 \rho^{5/3} \int_0^L C_n^2(h)\, dh, \qquad (l_o \ll \rho \ll \sqrt{\lambda L}), \tag{8.34}$$

$$[D_\varphi(\rho)]_{sp} = \frac{3}{8} 2.91 k^2 \rho^{5/3} \int_0^L C_n^2(h)\, dh, \qquad (\rho > \sqrt{\lambda L}). \tag{8.35}$$

The initial expression for the structure function evaluation in a SAR is Eq. (8.35), because there is the relation

$$\rho = d_e > \sqrt{\lambda L}.$$

The $C_n^2(h)$ function was shown above to be given by

$$C_n^2(h) = C_{n0}^2 \exp\left(-\frac{h}{h_0}\right).$$

As a result, we have the formula

$$\int C_n^2(h)\, dh = C_{n0}^2 h_0 \left[1 - \exp\left(-\frac{L}{h_0}\right)\right],$$

where $L = h_t \operatorname{cosec} \theta$, θ is the angle between the wave propagation direction and the skyline, and h_t is the total altitude of the turbulent stratum.

A synthetic aperture is characterised by the equality $\rho = d_e$, where d_e is the equivalent base at h_t. It follows from Fig. 8.4 that

$$d_e = \frac{L_s h_t \operatorname{cosec} \vartheta}{R_o} = \frac{L_s h_t}{H}.$$

Thus, we eventually get the relation

$$D_\varphi(\rho) = \beta_0 \left(\frac{2\pi}{\lambda}\right)^2 \left(\frac{L_s h_t}{H}\right)^{5/3} C_{n0} h_0 \left[1 - \exp\left(-\frac{h_t \operatorname{cosec} \vartheta}{h_0}\right)\right], \tag{8.36}$$

where $L_s = \bar{V} T_s$, T_s is the synthesis time, $\vec{V}$ is the track velocity of the radar carrier and $\beta_0 = 1.09$.

Equation (8.36) also allows finding the standard deviation of the phase difference fluctuations at the synthetic aperture ends:

$$\sigma_\varphi(\rho) = \sqrt{D_\varphi(\rho)}. \tag{8.37}$$

We shall now examine how phase errors due to tropospheric turbulence affect the resolution limit and optimal length of a synthetic aperture. W. Brown and Y. Riordan [23] have calculated both parameters for the case of phase errors, with the structure function obeying a power law. It was stated that the phase difference $[\varphi(r+\rho) - \varphi(r)]$ has a Gaussian distribution, and this is supported experimentally. For the above type of phase errors, the expression for the aperture resolution along the track is found to be

$$\rho_x = \frac{\lambda R}{4\pi\rho_0} \tag{8.38}$$

with $\rho_0 = 0.985 b$. The quantity b is to be calculated from the equation for the structure function of a phase error:

$$D_\varphi(\rho) = b^n \rho^n, \quad n = 5/3. \tag{8.39}$$

Then Eqs (8.38) and (8.39) yield

$$\rho_x = \frac{\lambda R}{4\pi} \frac{[D_\varphi(\rho)]^{3/5}}{\rho}. \tag{8.40}$$

Using the equation for the structure function of a phase error (8.36) and $\rho = d_e$, we get

$$\rho_x = \lambda^{-1/5} R C_0 (C_{n0}^2)^{3/5} (h_0)^{3/5} \left\{ 1 - \exp\left(-\frac{h_t \operatorname{cosec} \vartheta}{h_0}\right) \right\}^{3/5}, \tag{8.41}$$

where $C_0 = $ const. This equation shows that ρ_x varies but slightly with λ and increases slowly with increasing λ.

The optimal synthetic aperture affected by a turbulent troposphere [23] can be found as

$$L_{\text{opt}} = \frac{13.4}{b}. \tag{8.42}$$

Then Eqs (8.42) and (8.39) give

$$L_{\text{opt}} = \frac{d_0 \lambda^{6/5}}{(C_{n0}^2)^{3/5} (h_0)^{3/5} \left\{ 1 - \exp\left(-\frac{h_t \operatorname{cosec} \vartheta}{h_0}\right) \right\}^{3/5}} \tag{8.43}$$

with $d_0 = $ const.

The mean square value of the phase error between the optimal aperture centre and its extremal point is

$$\sigma_\varphi = (D_\varphi(L_{\text{opt}}/2))^{1/2}, \tag{8.44}$$

where D_φ and L_{opt} are to be calculated from Eqs (8.36) and (8.43).

Some other methods for reducing propagation-induced phase error in coherent imaging systems were suggested in Reference 22 and 47.

8.3 A model of phase errors in a turbulent ionosphere

It was shown in the Appendix to Reference 114 that a good approximation for the structure function of phase fluctuations is the expression:

$$D(y) \cong C_{\delta\Phi}^2 |y|^{2\nu-1}, \qquad 0.5 < \nu < 1.5. \tag{8.45}$$

The phase structure constant $C_{\delta\Phi}^2$ is defined as

$$C_{\delta\Phi}^2 = \frac{C_p}{2\pi} \frac{2\Gamma(1.5 - \nu)}{\Gamma(\nu + 0.5)(2\nu - 1)2^{2\nu-1}}, \qquad 0.5 < \nu < 1.5, \tag{8.46}$$

where $C_p = r_e^2 \lambda^2 l_p C_S$, l_p is the path length of an electromagnetic wave in the ionosphere, r_e is the classical electron radius $r_e = 2.81 \times 10^{-15}$ m, λ is the transmitter wavelength, and C_S is the turbulence parameter in the ionosphere described by Eq. (8.26).

Using the phase screen model of Reference 116 and Eq. (8.46), one can show that the mean square value of the phase fluctuations along the path l_p is defined as

$$\langle \delta\Phi^2 \rangle = 2\sqrt{\pi} r_e^2 \lambda^2 l_p C_S G \frac{\chi_0^{-2\nu+1} \Gamma(\nu - 1/2)}{4\pi \Gamma(\nu + 1/2)}, \tag{8.47}$$

where the factor G was borrowed from the Appendix to Reference 113. This factor accounts for:

- the velocity of the scanning beam motion relative to electron density whirls (v_o),
- the geometrical parameter due to the electron density anisotropy (Ω),
- the effective velocity of the scanning beam across the earth surface (V_{ef}),
- the synthesised aperture length L_s.

The factor G is defined as

$$G = \Omega(V_{ef} L_s / v_o)^{p-1}. \tag{8.48}$$

The equations for Ω and V_{ef} can be found in Reference 113.

All the fundamental concepts of the model we have just discussed were developed by Rino, so we think this model should bear his name. It has been successfully employed to analyse the effects of ionospheric turbulence on communication and navigation device performance. But we also believe that this model can be useful for the estimation of aperture performance in whirls and their effect on the azimuth ambiguity function. The latter is important because one can then evaluate the aperture resolution errors.

8.4 Evaluation of image quality[1]

Synthetic apertures were primarily designed for obtaining images to be used by a human operator to solve research and applied problems. It is natural that the evaluation of aperture performance should largely be based on the analysis of image characteristics. To do so, one needs to have at one's disposal appropriate criteria for a quantitative description of the performance characteristics of a particular type of aperture to be able to compare them with those of other apertures and to suggest appropriate improvements.

At present, there is no generally accepted criterion for evaluation of aperture performance or image quality, though there have been some attempts made along this line [99]. Difficulties involved in developing a reliable criterion are due not only to the complex design and random behaviour of a synthetic aperture but also to the diversity of their applications (e.g. a great variety of target aspect angles at which imaging is made). Normally, potential characteristics or some individual parameters are used as criteria for the evaluation of aperture performance.

8.4.1 Potential SAR characteristics

SAR designers and researchers often use the so-called potential characteristics, since they describe the aperture response to an echo signal from a point scatterer and do not contain micronavigation noise [53]. The following parameters may be referred to as potential characteristics. We shall mostly list characteristics of apertures using a digital signal processing and a digital image reconstruction.

1. The major lobe width of a synthetic antenna pattern (SAP) characterises the potential resolving power of an aperture in azimuth ρ_β. This parameter is determined by the width of the aperture response to a point target at zero noise. In practice, a 3 dB SAP width is most often used as a criterion for evaluation of a potential resolution, but there are other approaches, too. The potential resolution is usually evaluated with a uniform weighting function $H(t) \equiv 1$ to get

$$\rho_\beta \approx \lambda/(2L \sin \gamma), \tag{8.49}$$

   where L is the projection of the synthesis step onto the normal to the view line and γ is the incidence angle of microwave radiation. If the weighting function is non-uniform, the major lobe width becomes 1.2–2.5 times larger, depending on the type of the weighting function.

2. The integral level of side lobes

$$b_i = \left(\int_{-\pi}^{\pi} I^2(\beta) \, d\beta - \int_{-\rho_\beta/2}^{\rho_\beta/2} I^2(\beta) \, d\beta \right) \Big/ \int_{-\pi}^{\pi} I^2(\beta) \, d\beta \tag{8.50}$$

---

[1] Section 8.4 was written by E. F. Tolstov and A. S. Bogachev.

Table 8.1 The main characteristics of the synthetic aperture pattern

Type of weighting function	Relative SAP width	b_i	$20\lg(b_m)$
Uniform	1.0	0.0705	−13.3
Parabolic	1.3	—	−20.6
Henning's	1.6	0.0103	−32.0
Hamming's	1.45	0.178	−42.0

characterises the maximum SAP relative to the background created by the side lobes.

3. The maximum side lobe level is

$$b_m = I_{ms}/I_m, \qquad (8.51)$$

where I_{ms} and I_m are the maximum side and major lobe senses, respectively. This parameter is effective in sensing microwave-contrast targets against a weakly reflecting background. The integral and maximum senses of the side lobes, as well as the major lobe width, vary with the weighting function used in the SAR (Table 8.1). The relative width in the Table is the SAP width normalised to that for a uniform weighting function.

4. The azimuthal sample characteristic is

$$k_a = \rho_\beta/\rho_\Delta, \qquad (8.52)$$

where ρ_Δ is a step between the azimuthal counts of an image digital signal. According to the theorem of samples, the sample characteristic must meet the condition $\rho_\Delta < \rho_\beta$.

This parameter denotes the number of digital signal counts per azimuthal resolution element and describes the radar capability to reconstruct an image. The larger the sample characteristic, the greater the image contrast. However, a larger coefficient entails a greater complexity of the image reconstruction design. The optimal value of this parameter is taken to be $k_a = 1.2$.

5. Image stability characterises the ability of an image digital reconstruction device to sense and count the relative positions of partial frame centres and to provide the proper scale over all the sample characteristics when partial frames are matched and superimposed.

6. The gain in the signal-to-noise ratio in coherent and incoherent integration is calculated from the variations of this parameter at the processor output. It is assumed that the echo and image signals are integrated linearly in both coherent [17] and incoherent integration [59], whereas noise is integrated in quadratures. Therefore, the total gain in the signal-to-noise ratio K_g is

$$K_g = \sqrt{Nn}, \qquad (8.53)$$

where n is the number of echo counts over a synthesis step in one range channel and N is the number of incoherently integrated partial frames.

In real flight conditions, the actual aperture characteristics differ from the potential ones. The reason for this is the noise from processing and micronavigation devices, as well as the limitations of imaging systems.

8.4.2 Radar characteristics determined from images

The real performance characteristics of a radar system are evaluated from the results of a statistical processing of image parameters registered during experimental flights over a test ground (of the type of Willcox Playa in the United States). The radar characteristics to be found experimentally are usually as follows.

1. The realistic aperture sharpness is taken to be the minimal distance between two corner reflectors discernible on an image along the respective coordinate, if the reflectors produce pulses of equal intensity and if the power of the reflected signals is much greater than the noise. Note that the sharpness evaluation is affected by the sample characteristic that can normally be varied by the operator during the test.
2. The intensity of speckle noise on an image is defined as the ratio of the standard deviation to the mean signal intensity on an image for a statistically uniform area on the earth. The speckle arises from the presence of numerous point scatterers in a resolution element, which have an approximately identical radar cross section (RCS) and are produced by re-emission of the antenna pattern of random geometry. The speckle effect can be reduced by filtering or by incoherent integration of several independent images of the same region on the earth. Independent images can be obtained at different radiation frequencies, polarisations or aspect ratios. Depending on the SAR application, the number of such images varies from 3 to 4 for military applications to 70 for resources survey tasks.
3. The dark level on an image is an average intensity of a signal from a region of the lowest reflectivity. Sometimes, the dark level is taken to be the average image intensity with a zero echo signal at the input (the noise dark level). This parameter is related to the side lobe size in the synthesised antenna pattern and to the processing noise.
4. The dynamic range is defined as the maximum-to-minimum signal intensity ratio on an image. It depends on the design of the transmitter–receiver unit, the processor characteristics, the receiver gain control, etc.
5. The contrast of adjacent samples is found as the ratio of the maximum signal intensity from a point target (much above the noise level) to the average intensity of the adjacent samples. This parameter characterises the SAR ability to reconstruct the maximum space frequency on an image.
6. The mean image power is a parameter affected not only by the transmitter power, the antenna gain, the receiver sensitivity and the signal-to-noise gain at the processor output, but also by the post-processing before a signal is displayed (especially, at the stage of defining its minimum threshold).

176 Radar imaging and holography

7. The intrinsic aperture noise level is the mean image signal level when there is only noise at the aperture input and its gain corresponds to the mean image signal. This parameter covers the total effect of the aperture noise during the synthesis.
8. The radar swath width is determined by the screen parameters (the number of lines and the number of pixels in a line) and by the discretisation step in range and azimuth. An acceptable number of image pixels on a screen normally varies from 512 × 512 to 1024 × 1024.
9. Geometrical distortions of an image are defined as the standard deviation of the positions of reference scatterers relative to their actual positions. The central reference mark is superimposed with the real reference. The standard deviation value is affected by the range, the view angle, altitude, the distance between the reference and the image centre, as well as by the imaging time.
10. The imaging time is an important parameter of an aperture operating in real time. A typical test ground for the study of aperture characteristics is a statistically uniform surface with three-edge corner reflectors (Fig. 8.5) arranged at different distances from each other (for evaluation of the aperture sharpness). The reflectors possess different reflectivities, so one can measure the dynamic range of the system. In addition to a uniform background, a test ground usually includes some common objects such as roads, fields, smooth surfaces, railway roads, etc.

In order to understand better the difference between the potential and real characteristics of a synthetic aperture and a SAR as a whole, we shall make use of test results with digital image reconstruction (the AN/APQ-102A modification) [53]. Its potential resolution was 12.2 m along the azimuth and range coordinates. The discretisation step for evaluation of a real azimuthal resolution was taken to be 3.04 m. Figure 8.6 shows an azimuthal signal from two corner reflectors. When the valley

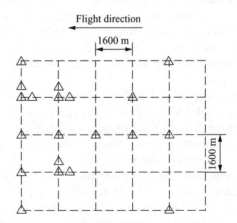

Figure 8.5 A schematic test ground with corner reflectors for investigation of SAR performance

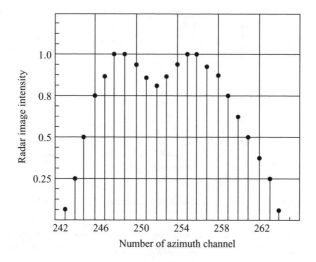

Figure 8.6   A 1D SAR image of two corner reflectors

between their images was 2 dB, the azimuthal resolution was found to be 21.28 m, or 7 pixels in an image line.

Part of the test ground image was obtained by a 14-fold incoherent integration with the mean signal value of 0.671 and a standard deviation of 0.201. The evaluated speckle was found to be 0.3, which is a sufficiently low level.

The dark level was typically 23 dB of the grey-level value. Hence, the SAR dynamic range is 33 dB, with the contrast of adjacent samples being 2.8 or 4.5 dB. For a synthetic aperture with strongly suppressed side lobes, this parameter was 6–10 dB. The large standard deviation in this case is due to the use of corner reflectors with a large RCS.

The dynamic range is estimated from these data to be 33 dB, with the contrast of adjacent samples being 2.8 or 4.5 dB. For a synthetic aperture with strongly suppressed sidelobes, this parameter is 6–10 dB. The large mean square value of the image signals is due to the application of corner reflectors with a large RCS.

Figure 8.7 shows a histogram of the noise distribution at the aperture output, and one may suggest that the probability density has a Rayleigh pattern. The mean value of 0.21 was taken to be the dark level. One of the dark regions exhibits a Rayleigh distribution with a mean value of 0.42. A screen with 384 × 360 pixels covered a view zone of 4.8 × 4.5 km. The errors in the measurement of the range positions of the corner reflectors were 14 km and 18 m at a distance of 1600 m from the image centre, whereas the radar was at 14.5 km from it. The azimuth measurement error was ∼50 m under the same conditions.

8.4.3   Integral evaluation of image quality

The authors of Reference 99 have suggested a method of integral evaluation of radar images. With this method one can compare images and establish a certain standard

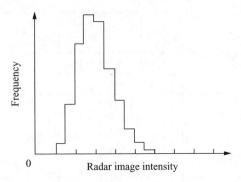

Figure 8.7 A histogram of the noise distribution in a SAR receiver

for the transformation of resolution to the number of incoherent integrations or to a parameter related to the dynamic range of an image signal. It is shown that the interpretability, or the operator's ability to interpret an image, U, is related to the SGL volume V as

$$U = U_0 \exp(-V/V_c), \tag{8.54}$$

where U_0 is the maximum image interpretability and V_c is the critical grey-level resolution.

It has been found empirically that the interpretability is related to the grey-level volume defined as

$$V = p_a p_r p_g, \tag{8.55}$$

where p_a, p_r are the linear resolutions in azimuth and range, respectively, and p_g is the grey-level resolution (in half-tones). The new image parameter – grey-level resolution – can be expressed as the ratio of a level a signal exceeds in 90 per cent of cases to that in 10 per cent of cases for independent samples. This parameter can be found from the formula:

$$p_g \approx (\sqrt{N} + 1.282)/(\sqrt{N} - 1.282). \tag{8.56}$$

However, the calculated value differs noticeably from the measurements made at $N < 4$ (Fig. 8.8). The experimental interpretability scale ranged from 0 for an uninterpretable image to 4 for a fully interpretable one. Therefore, the maximum interpretability U_0 should be 4.

The authors of Reference 1 have obtained a more complex equation for p_g

$$p_g = 10 \lg \frac{1 + (N/e) \sum_{k=1}^{N} 3k[(N-k)!N^{k+1}t]^{-1}}{1 - (N/e) \sum_{k=1}^{N} 3k[(N-k)!N^{k+1}]^{-1}}.$$

Where $e = 2.78$, this result, however, is based on the information theory and additionally takes into account properties of photointerpreter's visual analyser. It was discovered that according to the criterion of the maximum image information capacity $N = 2$ is optimal.

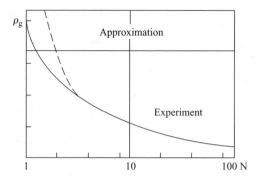

Figure 8.8 *The grey-level (half-tone) resolution versus the number of incoherently integrated frames N*

An important experimental finding was the critical volume V_c – for a single frame synthesised by the aperture ($N = 1$). For the majority of frames, the length per square resolution element in the case of a 37 per cent interpretability was found to be 9.14 m. Such objects were vegetation and urban areas, low-contrast regions, communication lines, city and country roads, etc. Exceptions were the boundaries of water bodies and vegetation covers showing a 37 per cent interpretability even at the lowest linear resolution in azimuth and range (13.72 m). Since the grey-level resolution at $N = 1$ (Fig. 8.8) is 22, it is easy to find the critical volume:

$$V_c = p_a p_r p_g \approx 9.14^2 \times 22 \sim 1850. \tag{8.57}$$

With this, the final interpretability expression takes the form:

$$U = 4\exp\{-p_a p_r p_g/1850\}. \tag{8.58}$$

Note that the calculation of the critical volume used the linear resolution of 9.14 m. Figure 8.9 shows the interpretability plotted against the linear resolution $p_a = p_r = p$ for different numbers of incoherent integrations.

When analysing the plots in Fig. 8.9, one should bear in mind that both the measurements and the calculations were based on some *a priori* assumptions. For example, the half-tone scale was chosen on the assumption that a photograph had the maximum interpretability and that it had an infinite number of incoherent integrations ($N = \infty$) and the half-tone resolution p_g – (Fig. 8.8). An image synthesised without incoherent integrations ($N = 1$) was thought to have the poorest half-tone resolution, but the resolution was to be finite ($p_g < \infty$), since the image preserved some, though very low, interpretability. It was established experimentally that the poorest half-tone resolution was equal to 22 (Fig. 8.8).

The interpretability was evaluated by three qualified and experienced interpreters of radar and optical images, using the four-level scale (from 0 to 4) mentioned above. The interpreters worked with prints of 20.32 cm × 25.40 cm in size. The resolution elements varied in shape from square to rectangular (with the side ratio of 1:10) and in the number of incoherent integrations varying from 1 to ∞. All the experiments

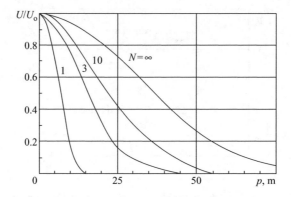

Figure 8.9 The dependence of the image interpretability on the resolution versus linear resolution $p_a = p_r = p$

were carried out using a quadratic detector because the detection was performed on a quadratic film. It can be demonstrated theoretically, however, that experimental data can also be useful in linear detection of image signals if the half-tone resolution is calculated by another approximate formula:

$$p_{gl} \approx (\sqrt{N} + 0.6175)/(\sqrt{N} - 0.6175). \tag{8.59}$$

The major result of this series of investigations [99] was the experimental support of the idea that image interpretability depended only on the half-tone volume resolution, or on the product of the azimuthal, range and half-tone resolutions. Therefore, this parameter varies with the area rather than the shape of a resolution element (square or rectangular). On the other hand, it depends on the resolution element area and the number of incoherent integrations. So one can make a compromise when choosing the resolution in azimuth p_a, in range p_r and in half-tones p_g [99]. Identical interpretabilities can be achieved by using different combinations of these parameters. This conclusion proved to be quite unexpected and may play an important role in solving some applied problems when one has to choose between the complexity and the cost of aperture processing techniques.

Indeed, if this conclusion is correct, it is worth making an effort to achieve a high image interpretability by improving low-cost resolutions. To illustrate, a higher range resolution and an incoherent integration in spaceborne SARs can be achieved in a simpler way than a higher azimuthal resolution. For example, one can fix the azimuthal resolution but improve the range resolution or increase the number of incoherent integrations.

We shall give a good example to illustrate the effectiveness of resolution redistribution with reference to a side-looking synthetic aperture. In this type of aperture, the azimuthal resolution depends linearly on the number of incoherent integrations N:

$$p_a(N) = \lambda r_o N/2L_m = p_o N, \tag{8.60}$$

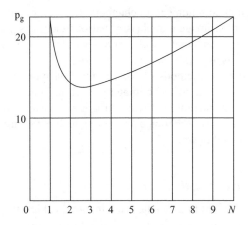

Figure 8.10 *The dependence of the half-tone resolution on the number of incoherent integrations over the total real antenna pattern*

where λ is the wavelength, r_o is the oblique range, L_m is the maximum possible length of the aperture, and $p_o = \lambda r_o / (2L_m)$ is the best aperture resolution. If we now fix the range resolution, the minimum product of $p_a N p_g$ will show the optimal combination of azimuthal resolution and incoherent integration (Fig. 8.10). This optimum is found to lie at $N = 3$; hence, $p_a = 3p_o$.

The integral criterion for image evaluation from the half-tone volume resolution is convenient and relatively simple. But when using it in practice, one should bear in mind that the available amount of statistical data is insufficient, so the estimations of image quality may be quite subjective.

8.5 Speckle noise and its suppression

Synthetic aperture radar remote sensing of the earth is becoming increasingly popular in many areas of human activity (Chapter 9.1). The analysis of images may be made in terms of a qualitative or quantitative approach [2].

A qualitative analysis is largely made by conventional methods of visual interpretation of aerial photography, combined with the researcher's knowledge and experience. Although radar images have much in common with aerophotographs (Chapter 1), the physical mechanisms of their synthesis set limits on the applicability of interpretation methods elaborated for optical imagery. Additional difficulties arise from the presence of speckle noise.

A quantitive analysis is based on the measurement of target characteristics for various backgrounds and objects [2], followed by computerised processing of video information. The latter is normally used to solve the following tasks. One often has to improve image quality and interpretation procedures at the pre-processing stage, which includes various corrections, noise reduction, contrast enhancement, highlighting contours, etc. It may also be necessary to compress and code images to be transmitted through communication channels. Besides, one may have to identify some of the items on an image and classify various elements present on it. This is usually done by image segmentation, cluster analysis and so on. Obviously, this kind of image subdivision is always somewhat arbitrary.

Here we shall discuss methods of solving the first type of task with emphasis on those techniques specific to radar imagery, such as speckle suppression. Some others, like geometrical and radiometrical correction, have already been dealt with in the literature [2,31]. Some of the image processing techniques are quite versatile and have also been discussed in detail [2].

8.5.1 Structure and statistical characteristics of speckle

There has been much effort to understand the image speckle structure. The available publications on this subject can be classified into two groups as for the specific problems being tackled. The more extensive group covers work on speckle as a noise, suggesting various ways of its filtering. The other group includes publications on useful properties of speckle, in particular, on the possibility to derive from it information about the area of interest. Naturally, there are problems in each trend that remain poorly understood. A feature common to all the publications is the description of statistical characteristics of speckle.

Let us consider the statistical characteristics of an echo signal in terms of a general reflection model when a resolution element contains many echo signals from different point scatterers. The signals are random, independent and have about the same intensity. Then the total signal represents a Gaussian random quantity and its amplitude has a Rayleigh pattern. This kind of reflection model is often termed the Rayleigh model. When a synthetic aperture changes its position relative to a target, the intensity fluctuations of the total echo signal give rise to a characteristic speckle pattern on an image. Clearly, the intensity I of individual pixels will obey the exponential law of the probability density distribution:

$$p_I(x) = \frac{1}{2\sigma_0^2} \exp\left(-\frac{x}{2\sigma_0^2}\right) \tag{8.61}$$

with the mean value of $\bar{I} = 2\sigma_0^2$ and the dispersion $\sigma_I^2 = 4\sigma_0^4$, while the phase θ of the image pixels is equiprobable in the range from $-\pi$ to $+\pi$.

Another reflection model is applied when a resolution element has one bright point together with other point scatterers, such that the total echo signal contains one dominant signal of much higher intensity along with many random independent signals of nearly the same lower intensity. Then the amplitude of the total signal is described by the Rice distribution, or by a generalised Rayleigh distribution. This kind of model is called the Rice reflection model.

The distribution of the intensity probability density at single pixels is

$$p_I(x) = \frac{1}{2\sigma_0^2} \exp\left(-\frac{x - s_0}{2\sigma_0^2}\right) I_0\left(\frac{\sqrt{xs_0}}{\sigma_0^2}\right) \tag{8.62}$$

with the mean value of $\bar{I} = 2\sigma_0^2 + s_0$ and the dispersion $\sigma_I^2 = 4\sigma_0^4(1+2r)$, where s_0 is the square amplitude of the highest intensity component of the signal $r = s_0/(2\sigma_0^2)$, $I_0(\cdot)$ is a modified zero-order Bessel function of the first kind, and the distribution of the phase probability density is

$$p_\theta(x) = \frac{1}{2\pi} \exp\left(-\frac{a^2}{2}\right) + a\frac{\cos x}{\sqrt{2\pi}} \Phi(a \cos x) \exp\left(-a^2 \sin^2 \frac{x}{2}\right), \tag{8.63}$$

where

$$a = \sqrt{s_0}/\sigma_0, \qquad \Phi(t) = \frac{1}{2} \int_{-\infty}^{t} \exp\left(-\frac{\tau^2}{2}\right) d\tau$$

is the Laplace function.

Since the signal-to-noise ratio is the ratio of the mean intensity $\bar{I}$ to the standard deviation σ_I equal to 1 and $(1+r)/\sqrt{1+2r}$ for the Rayleigh and Rice models, respectively, the intensity fluctuation amplitude in the speckle structure is commensurable with the useful signal intensity for a complex target at $r \approx 1$. For this reason, images of such targets have a well-pronounced speckle structure. Since it is difficult to analyse an echo signal from a target with the Rice reflection, most authors discuss targets with that of the Rayleigh reflection.

It is worth noting that the above expressions for the probability density distribution in the case of a uniform and isotropic background are valid for both an ensemble of images at each resolution element and a single image over a multiplicity of resolution elements. For a non-uniform background, however, these expressions are valid only for an ensemble of image realisations.

When the N number of independent images of the same earth area are summed up, the probability density distribution of the speckle structure takes the form:

$$p_I(x) = \frac{x^{N-1} \exp(-(x/2\sigma_0^2))}{(2\sigma_0^2)^N \Gamma(N)} \tag{8.64}$$

with the value of $\bar{I} = 2N\sigma_0^2$ and the dispersion $\sigma_I^2 = 4N\sigma_0^4$, where $\Gamma(\cdot)$ is the gamma-function described as $\Gamma(N) = (N-1)!$ for integer N. In this case, the signal-to-noise ratio is $\sqrt{N}$. The probability density distribution in Eq. (8.64) corresponds to the gamma-distribution with the parameters equal to N and $1/(2\sigma_0^2)$, or to χ^2-distribution with $2N$ degrees of freedom at $\sigma_0^2 = 1$. A general expression for the initial moments of distribution (8.64) has the form:

$$M_k^N = [(N+K-1)!/(N-1)!](2\sigma_0^2)^k,$$

where M_k^N is the kth initial moment.

Reference 2 presents the fluctuation spectrum of the speckle amplitude and its autocorrelation function. Suppose a point scatterer is described by the Dirac δ-function and $F(k)$ is the transfer function of a synthetic aperture, where $k = 2\pi/\lambda$ is the wave number and λ is the wavelength of the echo signal. Then the amplitude spectrum of the echo signal from a point scatterer located at a point with the coordinate x relative to the SAR carrier track is $F' = (1/2)F(k)\exp(jxk)$. For randomly arranged point scatterers, the signal received by the aperture is defined as

$$F'(k) = \frac{1}{2\pi}F(k)\sum_{l=1}^{L}\exp(jx_l k).$$

At $L \to \infty$, the speckle power density spectrum can be determined within the accuracy of a constant factor:

$$S(k) = \overline{|F'(k)|^2} = |F(k)|^2.$$

In other words, it is unambiguously dependent on the aperture transfer function. The autocorrelation function of speckle is related to its spectral power density by a Fourier transform. Therefore, the speckle autocorrelation function can be used to find the aperture impulse response directly.

The statistical characteristics of speckle for a background represented as an array of randomly moving point scatterers are considered in Reference 2. It is shown that the concept of spatial resolution has no sense if the phase fluctuations of signals from the point scatterers are large (the phase changes by 2π several times during the synthesis).

8.5.2 Speckle suppression

The available methods of suppression or smoothing out of image speckle can be subdivided into two groups. Some methods are based on the averaging of several independent images of the same background. This group is not large but these methods have been extensively used owing to their relative simplicity. The other group of methods is much larger and includes so-called aposterior procedures when speckle is suppressed by spatial filtering.

Independent images of the same earth area can be obtained in different ways based on a common principle of image segmentation with respect to a particular parameter, for example, the Doppler frequency, the carrier frequency or polarisation (i.e. sensing a background at different polarisations of probing radiation). The first technique is known as a multibeam processing and it is most commonly used in practice [99]. A specific feature of multibeam processing is a proportional decrease of the aperture sharpness in track range when the Doppler frequency band is subdivided into N identical non-overlapping subbands. The specificity of speckle suppression procedures is that the signal-to-noise ratio increases by a factor of $\sqrt{N}$ if N independent images are averaged.

The methods of the first group can use other procedures, for example, median filtering [2], in addition to the averaging of N independent images.

A wide application of aposterior techniques is primarily due to a rapid development of image processing technology. The lack of an adequate model of speckle structure and useful signal makes it difficult to design effective algorithms for speckle suppression. Until recently, nearly all researchers working on speckle problems have regarded speckle as a multiplicative noise to a useful signal. However, there are more complex models. The authors consider the possibility of employing Wiener's and Calman's filtering algorithms, homomorphic processing and various heuristic techniques to suppress speckle.

However, a lack of objective criteria for evaluation of image quality by visual perception creates additional difficulties. For this reason, nearly all the researchers cited below compare the processing results with expertise, which makes a comparative analysis of the suggested algorithms quite problematic.

The first attempts to suppress speckle by aposterior techniques used the Wiener filtering algorithm which varies with the signal [2]. The workers analysed an additive, signal-modelled noise approach and a multiplicative noise model. In the former, a distorted image is described by the expression:

$$z(x,y) = s(x,y) * h(x,y) + f[s(x,y) * h(x,y)]n(x,y), \qquad (8.65)$$

where $h(x,y)$ is the space impulse response, f is commonly a non-linear function and $n(x,y)$ is signal $s(x,y)$ independent noise. By introducing the designations $n'(x,y) = s'(x,y) \times n(x,y)$ and $s'(x,y) = f[s(x,y) * h(x,y)]$, we transform Eq. (8.65) to

$$z(x,y) = s(x,y) * h(x,y) + n'(x,y).$$

In the second noise model, an image is described as

$$z(x,y) = n(x,y)[s(x,y) * h(x,y)], \qquad (8.66)$$

where $n(x,y)$ is signal-independent multiplicative noise. The Wiener's filter has the transfer function $M(\mu,v) = \Phi_{zs}(\mu,v)/\Phi_{zz}(\mu,v)$ and minimises the standard deviation of the filtering, provided that $z(x,y)$ and $s(x,y)$ are wideband spatially uniform random fields, Φ_{zs} and Φ_{zz} are the respective power density spectra. With Eq. (8.65), the first noise model gives the following transfer function of a Wiener's filter:

$$M_1(\mu,v) = \frac{\Phi_{ss}(\mu,v)H^*(\mu,v)}{\Phi_{ss}(\mu,v)|H(\mu,v)|^2 + \Phi_{s's'}(\mu,v) * \Phi_{nn}(\mu,v)} \qquad (8.67)$$

on the assumption of $\overline{n(x,y)} = 0$. Here $n(x,y)$ is statistically independent of $s(x,y)$, $s'(x,y)$ is a uniform wideband field, and $H(\mu,v) = F[h(x,y)]$ is the system's transfer function. At $f[s(x,y) * h(x,y)] = s(x,y) * h(x,y)$, we have $\Phi_{s's'}(\mu,v) = \Phi_{ss}(\mu,v)|H(\mu,v)|^2$, and Eq. (8.67) can be re-written as

$$M_1(\mu,v) = \frac{\Phi_{ss}(\mu,v)H^*(\mu,v)}{\Phi_{ss}(\mu,v)|H(\mu,v)|^2 + [\Phi_{ss}(\mu,v)|H(\mu,v)|^2] * \Phi_{nn}(\mu,v)}. \qquad (8.68)$$

If the noise is uniform, wideband and signal-independent, the transfer function of a Wiener's filter in the second model will be

$$M_2(\mu,v) = \frac{\overline{n}\Phi_{ss}(\mu,v)H^*(\mu,v)}{\Phi_{nn}(\mu,v) * [\Phi_{ss}(\mu,v)|H(\mu,v)|^2]}. \qquad (8.69)$$

It is clear from (8.69) that at $\bar{n}(x, y) = 0$ the filter transfer function is $M_2(\mu, \nu) = 0$. Suppose we have $n_1(x, y) = n(x, y) - \bar{n}$, then

$$M_2(\mu, \nu) = \frac{\Phi_{ss}(\mu, \nu) H^*(\mu, \nu) / \bar{n}}{\Phi_{ss}(\mu, \nu) |H(\mu, \nu)|^2 + (1/\bar{n}^2) \Phi_{n_1 n_1}(\mu, \nu) \otimes [\Phi_{ss}(\mu, \nu) |H(\mu, \nu)|^2]}. \quad (8.70)$$

Obviously, at $n = 1$ filters with the transfer functions (8.68) and (8.70) are equivalent. Modelling has shown that a Wiener's filter for signal-dependent noise with the characteristics M_1 and M_2 is better than that for additive, signal-independent noise. But the essential limitations of the former are the need for a large amount of *a priori* information about the signal and the noise, as well as vast computations. Calman's filtering algorithms [2] suffer from similar disadvantages.

The possibility of a homomorphic image processing is discussed in Reference 2. A homomorphic processing is supposed to be any conversion of observable quantities if the signal fluctuations are transformed to additive and signal-independent noise. Within the multiplicative speckle model, Eq. (8.64) yields

$$p(I) = \frac{N^N}{\Gamma(N)\bar{I}} \left(\frac{I}{\bar{I}}\right)^{N-1} \exp\left(-\frac{NI}{\bar{I}}\right) \quad (8.71)$$

with $\sigma_I^2 = \bar{I}^2 / N^2$. Then the homomorphic transformation reduces to taking the logarithms. The distribution density of the quantity $D = \ln I$ is described as

$$p(D) = [N^N / \Gamma(N)] \exp[-N(D - D_o)] \exp\{-N \exp[-(D - D_o)]\} \quad (8.72)$$

with $D_o = \ln \bar{I}$. Practically, the distribution of signal-dependent noise is often approximated by a normal distribution with a signal-dependent dispersion. At any value of N, the approximation accuracy for the normal distribution (8.72) is greater than that for the distribution (8.71). The variable D can be processed by any algorithm available in the model of additive and signal-independent noise. It is pointed out in Reference 2 that the application of Wiener's filtering algorithm with a preliminary homomorphic processing of an image provides better results than a separate application of each algorithm.

The authors of Reference 2 believe that a homomorphic transformation is a reasonable alternative to image processing in signal-dependent noise. On the other hand, experience indicates that this does not give an essential advantage over heuristic methods to be discussed below. Moreover, the necessity to use both direct and inverse transformations increases the computation costs considerably.

There is another way of suppressing speckle noise – a local statistics technique [2]. Within the multiplicative speckle model, every element z_{ij} on an image is represented as the product of the signal s_{ij} and the noise n_{ij}. The noise has $\bar{n} = 1$ and the dispersion σ_n^2. On the assumption that the signal and the noise are independent, the authors have derived the expressions

$$\bar{z} = \bar{s}\bar{n} = \bar{s}$$

and

$$\sigma_z^2 = M[(sn - \bar{s}\bar{n})^2] = M[s^2]M[n^2] - \bar{s}^2\bar{n}^2.$$

If the signal intensity averaged over the processing window is constant, the expressions are

$$M[s^2] = \bar{s}^2 \quad \text{and} \quad \sigma_z^2 = \bar{s}^2(M[n^2] - \bar{n}^2) = \bar{s}^2\sigma_n^2 \text{ or } \sigma_n = \sigma_z/\bar{z}.$$

This model is consistent with the data obtained from the analysis of uniform surface imagery. The standard deviation σ_n is found to be about 0.28, which is due to a multi-beam processing and the use of other algorithms for improving images synthesised by the SAR SEASAT-A. Using the local statistics technique for a selected window (usually with 5×5 or 7×7 resolution elements), one can find the moving local average $\bar{z}$ and the dispersion σ^2. Then one gets

$$\bar{s} = \bar{z}/\bar{n}, \quad \sigma_s^2 = \frac{\sigma_z^2 + \bar{z}^2}{\sigma_n^2 + \bar{n}^2} - \bar{s}^2. \tag{8.73}$$

The expansion of z into a Taylor series with the account of the first-order terms only yields

$$z = \bar{n}s + \bar{s}(n - \bar{n}). \tag{8.74}$$

According to Eqs (8.73) and (8.74), the minimisation of the mean square error of speckle suppression leads to the following formula for $\hat{s}$:

$$\hat{s} = \bar{s} + k(z - \bar{n}\bar{s}) \tag{8.75}$$

with

$$k = \frac{\bar{n}\sigma_s^2}{\bar{s}^2\sigma_n^2 + \bar{n}^2\sigma_s^2}.$$

Then at $\bar{n} = 1$, one gets

$$\hat{s} = \bar{s} + k(z - \bar{s}), \quad k = \frac{\sigma_s^2}{\bar{s}\sigma_n^2 + \sigma_s^2}. \tag{8.76}$$

The heuristic algorithm derived from the local statistics approach is especially effective for speckle suppression on images of uniform and isotropic surfaces. It does not remove the contours of extended proper targets. This algorithm has provided good results when processing imagery from the SAR SEASAT-A. Its major advantages are simplicity and adaptive properties associated with the computation of the local statistics. It has, however, a serious limitation: it cannot predict the error behaviour during the speckle suppression. Besides, the necessity of computing the local average and, especially, the dispersion in a common 7×7 window considerably reduces the algorithm efficiency.

In order to decrease the computational costs inherent in local statistics algorithms, some workers have suggested using a sigma-filter. For a moving window of $(2m_1 + 1) \times (2m_2 + 1)$ in size (m_1 and m_2 are integer numbers) with the central resolution element z_{ij}, the signal $\hat{s}_{ij}$ is found from the formula:

$$\hat{s}_{ij} = \sum_{k=i-m_1}^{m_1+i} \sum_{l=j-m_2}^{m_2+j} \delta_{kl} z_{kl} \bigg/ \left[\sum_{k=i-m_1}^{m_1+i} \sum_{l=j-m_2}^{m_2+j} \delta_{kl} \right], \qquad (8.77)$$

where

$$\delta_{kl} = \begin{cases} 1, & \text{at } (1 - 2\sigma_n)z_{ij} \leq z_{kl} \leq (1 + 2\sigma_n)z_{ij}. \\ 0, & \text{otherwise}. \end{cases}$$

It is clear that a filter with the characteristic (8.77) will be more cost-effective than that with (8.76). A 11×11 window was used in Reference 2 to estimate σ_n. It was found that two passes of a sigma-filter were sufficient to get a satisfactory suppression of speckle noise without smearing the contours. When the number of passes was increased to four and more, the image was damaged.

The following modification of the sigma-filter was discussed in Reference 2 for filtering impulse noise together with speckle suppression. One chooses the threshold B. If the number of elements to be removed in accordance with Eq. (8.77) is smaller than or equal to the threshold B, the average of four neighbouring elements is ascribed to the estimated position of the moving window. The choice of a threshold is critical because it affects the contours. It is pointed out in this work that the threshold value for a 7×7 window should be less than 4 and for a 5×5 window less than 3. The use of a sigma-filter with a 11×11 window and then another sigma-filter with a 3×3 window at the threshold $B = 1$ proved to be most effective. A small window allows suppression of impulse noise in the vicinity of sharp contours. Other filter modifications are also possible. This type of filter was compared with a filter with the characteristic (8.76) and with a median and an averaged filter. It was concluded from the expertise that a sigma-filter provides better results. Its disadvantage is that one cannot estimate *a priori* the behaviour of the speckle suppression error. An important merit of this type of filter is its simplicity, a high computational efficiency and additive properties. These characteristics make the filter suitable for application in digital image processing in a real-time mode.

The local statistics method can also be implemented with a linear filter minimising the mean square error of the filtering. In addition to the algorithms described above, there is a large number of heuristic algorithms for speckle suppression. Among these are algorithms for median filtering, averaging over a moving window with various weighting functions, algorithms for a nonlinear transformation of the initial image, the reduction of an image histogram to a symmetric form, etc. Most heuristic algorithms are simple to use and have a fairly high computation efficiency but all of them possess a serious drawback – they practically ignore the specific process of SAR imaging: while suppressing noise, they partly suppress the useful signal. It is usually hard to estimate the speckle suppression error when using such algorithms.

To conclude, image processing covers a wide range of tasks and problems, many of which have not been dealt with in this chapter. Among these are the processing based on the properties of a human visual analyser, the criteria for image quality and image optimisation, quantitative evaluation of information contained in an image, etc. Due to a rapid development of cybernetics, information theory, iconics and computer science and practice, these areas of investigation are constantly trying new approaches. For example, they have tested some concepts of artificial intelligence in the processing of data on remote probing of the earth, the use of radar imagery as a database for visual interpretation and complexing of images obtained in different wavelength ranges. The results obtained from such studies can provide more information about the earth and other planets.

Chapter 9
Radar imaging application

9.1 The earth remote sensing[1]

9.1.1 Satellite SARs

Synthetic aperture radar imagery from satellites and aircraft has a high spatial resolution and is independent of light and clouds. Nearly real-time information and a comprehensive SAR image analysis is of importance not only for scientific studies, but also because it has a practical significance providing information for companies dealing with off-shore oil and gas exploration, deep-ocean mining, fishing, marine transportation, weather forecast, etc. [65]. In 1972 the NASA Office of Applications initiated the Earth and Oceans Dynamics Applications Program for the development of techniques of global monitoring of oceanographic phenomena and the design of an operational ocean dynamics monitoring system. Satellite SAR studies of the earth environment began in 1978, when the first series of images was obtained by the SEASAT during its 3 month's operation. This L-band horizontally polarised radar operated at a wavelength of 23 cm at an incidence angle of 20°. It was primarily designed for ocean wave imaging, although SAR imagery was also acquired over ice and terrestrial surfaces. It demonstrated the potential of satellite radar data in scientific and operative applications. The SEASAT data supported the notion that wind and wave conditions over the ocean could be measured from a satellite with an accuracy comparable to that achieved from surface platforms [5]. Various SAR instruments operating at different wavelengths, polarisations and incidence angles were mounted on bound of Space Shuttles (Table 9.1). In November 1981 and October 1984, the SIR-A and SIR-B radars, which used the SEASAT technology

---

[1] Sections 9.1.1 and 9.1.2 were written by V. Y. Alexandrov, O. M. Johannessen and S. Sandven, Nansen International Environmental and Remote Sensing Centre, St Petersburg, Russia Nansen Environmental and Remote Sensing Centre, Bergen, Norway. Section 9.1.3 was written by D. B. Akimov, Nansen International Environmental and Remote Sensing Centre, St Petersburg, Russia.

192 Radar imaging and holography

Table 9.1 Technical parameters of SARs borne by the SEASAT and Shuttle

Parameter	SAR				
	SEASAT	SIR-A	SIR-B	SIR-C/X	X-SAR
Orbit inclination (°)	108	38	57	57	57
Altitude (km)	800	260	225	225	225
Incidence angle (°)	20–26	47–53	15–60	20–55	20–55
Frequency (GHz)	1.28	1.28	1.28	1.25 and 5.3	9.6
Polarisation	HH	HH	HH	HH, VV, VH, HV	VV
Swath width (km)	100	50	30–60	15–90	15–45
Pixel size for four looks (m)	25 × 25	40 × 40	25	25	30 × (10 − 20)

Table 9.2 Parameters of the Almaz-1 SAR

Parameter	Value
Satellite altitude (km)	270–380
Orbit inclination (°)	72.7
Wavelength (cm)	9.6
Polarisation	HH
Radiometric resolution, one look (dB)	2–3
Swath width (km)	40
Spatial resolution, one look (m)	10–15

with the 23 cm wavelength and HH (Horizontal–Horizontal) polarisation, provided data targeted at land applications [77]. The SIR-C mission using a two-frequency multipolarisation SAR with a variable incidence angle, together with the X-band VV (Vertical–Vertical) SAR, operated in three flights during the period of 1994–1996. The SIR-C was of interest to ocean remote sensing, and its data were used to extend the understanding of radar backscatter from the ocean and SAR imaging of oceanographic processes [117].

The first USSR SAR mission started in July 1987 with a launch of the Cosmos-1870 satellite equipped with a S-band SAR. Its operation ended in July 1989 and was followed by the Almaz-1 satellite, which operated from May 1991 until October 1992 (Table 9.2). The raw data of 300 km long and 40 km wide stripes with a 10–15 m spatial resolution (one look) could be stored aboard and transmitted to a receiving ground station near Moscow as analogue radio holograms, with SAR images presented as photographic hard copies. Applications of SAR data included studies of various ocean phenomena and sea ice [36].

Table 9.3 The parameters of the ERS-1/2 satellites

Parameter	Value
Satellite altitude (km)	785
Orbit inclination (°)	98.52
Wavelength (cm)	5.66
Polarisation	VV
Angle of incidence (°)	20–26
Swath width (km)	100
Spatial resolution, three looks (m)	26 × 30

The first European Space Agency ERS-1 satellite with a C-band SAR aboard operated successfully from its launch in July 1991 until 1996 and provided a large amount of global and repeated observations of the environment. The focus was on ocean studies and sea ice monitoring [62,64]. In the high-resolution imaging mode, the ERS-1 SAR provides three-look, noise-reduced images with a spatial resolution of 26 m in range (across-track) and 30 m in azimuth (along-track) (Table 9.3). Because of the absence of onboard data storage, a network of ground receiving stations enabled a wide coverage by SAR images. ERS-2, a second satellite of this series, was launched in April 1995 and since mid-August 1995 both satellites operated in a tandem mode, when ERS-2 imaged the same area as ERS-1 one day later.

The RADARSAT launched by the Canadian Space Agency in November 1995 was the first SAR satellite with a clear operational objective to deliver data on various earth objects. Using the onboard data storage, it provides a much wider coverage than the ERS SAR [77]. Processed SAR data could be delivered to users within several hours after acquisition. The RADARSAT operates in the C-band and HH-polarisation, and in several imaging modes with different combinations of the swath width and resolution (Table 9.4). One of its main applications is sea ice monitoring [42].

The advanced SAR (ASAR) onboard the European Space Agency ENVISAT satellite, has been providing image acquisition since 2002 [43]. While its major parameters are similar to those of the RADARSAT, the ASAR can also operate at multipolarisation modes using two out of five polarisation combinations: VV, HH, VV/HH, HV/HH and VH/VV. The five major modes are: global, wide swath, image, alternating polarisation and wave modes (Table 9.5). In the image and alternating polarisation modes the ASAR gives high-resolution data (30 m and 3 look) in a relatively narrow swath (60–100 km), which can be located at different distances from the subsatellite track at the incidence angles from 15° to 45°. The alternating polarisation mode provides two versions of the same scene, at HH, VV and/or cross-polarisation. The wide swath mode provides a 420 km swath with a spatial resolution of 150 m and 12 looks. In the global monitoring mode, the ASAR continuously gives a 420 km swath with a spatial resolution of 1000 m and 8 looks.

Table 9.4 SAR imaging modes of the RADARSAT satellite

RADARSAT-1 modes with selective polarisation	Beam modes	Nominal swath width (km)	Incidence angles to left or right side (°)	Number of looks	Spatial resolution (approx.) (m)
Transmit H or V	Standard	100	20–50	1 × 4	25 × 28
Receive H or V	Wide	150	20–45	1 × 4	25 × 28
or (H and V)	Small incidence angle	170	10–20	1 × 4	40 × 28
	High incidence angle	70	50–60	1 × 4	20 × 28
	Fine	50	37–48	1 × 1	10 × 9
	ScanSAR wide	500	20–50	4 × 2	100 × 100
	ScanSAR narrow	300	20–46	2 × 2	50 × 50

Table 9.5 The ENVISAT ASAR operation modes

Operation mode parameter	Image mode	Alternating/ cross-polarisation	Wide swath mode	Global monitoring	Wave mode
Polarisation	VV or HH	VV/HH, HH/HV or VV/VH	VV or HH	VV or HH	VV or HH
Spatial resolution (along-track and across-track) (m)	28 × 28	29 × 30	150 × 150	950 × 980	28 × 30
Radiometric resolution (dB)	1.5	2.5	1.5–1.7	1.4	1.5
Swath width (km)	Up to 100 (seven subswaths)	Up to 100 (seven subswaths)	400 (five subswaths)	≥400 (five subswaths)	5 KM (vignette seven subswaths)
Incidence angle (°)	15–45	15–45			15–45

At present, SAR data from the ERS, RADARSAT and ENVISAT satellites are widely used in earth observations and monitoring of various natural objects and phenomena. With its fine-scale resolution, a SAR is capable of observing a number of unique oceanic phenomena [117]. These include wind and waves [46,75], ocean circulation [63], internal waves [33], oil spills [40,41], shallow sea bathymetry [6], etc. Imaging radars are also used in a number of land applications, such as the study of soil moisture [84], forestry [97] and the studying and monitoring of urban areas [135]. The use of satellite SAR data for monitoring the Arctic sea ice is briefly discribed below.

9.1.2 SAR sea ice monitoring in the Arctic

9.1.2.1 The use of satellite SAR for sea ice monitoring

The use of visible images for sea ice monitoring in the Arctic is limited by light in winter, while the cloud cover precludes sea ice observations in the visible and infrared ranges during approximately 80 per cent of time in summer [18,37,123]. Therefore, the development of remote radar sensing is essential for the polar regions. The first satellite SAR images were acquired by the SEASAT satellite which produced over 100 passes over the Beaufort Sea on nearly a daily basis for the analysis of sea ice motion and changes in the ice distribution. The SIR-B SAR gave data on the Antarctic sea ice margin for October 1984 [45]. Several SAR surveys were made over the Antarctic and Arctic with the Kosmos-1870 and Almaz-1 SARs in spite of the fact that the satellite orbits precluded coverage of the high-latitude northern and southern regions. The Almaz-1 SAR data were used to support an emergency operation in the Antarctic, when the research vessel Mikhail Somov got stuck in the ice. During this operation, it was possible to detect icebergs and estimate their size, as well as to derive several sea ice parameters, such as the ice extent, the boundaries of stable and unstable fast ice, the ice types (nilas, young and first-year ice), prevailing ice forms, ridges and areas of strongly deformed ice [3].

The SAR images obtained from ERS-1/2 were used in a number of sea ice studies in the Arctic, Antarctic and in the ice-covered seas in different parts of the World Ocean [48,68,76,93,120]. The ERS-1 SAR proved to be a very powerful instrument for sea ice observations. Although the ERS satellite was not designed for operational service, the data were applied in sea ice monitoring in the United States, Canada, Finland and several other countries [18,27].

With the launch of the Canadian RADARSAT in 1995, the first satellite with operational ice monitoring as a prime objective, ice monitoring in the United States, Canada, Greenland, Norway, Finland, Sweden and some other countries entered a new era. The ScanSAR mode with a swath of 450 km wide and with a 100-m resolution at 8 looks allows daily mapping of the whole polar region north of 70°N, and it is used for operational ice services in the Canadian Arctic, the Greenland Sea, the Baltic Sea and other areas with ice [18,48,111]. With a systematic acquisition of ScanSAR images over large Arctic sea ice areas and the use of the RADARSAT geophysical processor, it was possible to estimate the sea ice motion, deformation and thickness from sequential imagery for several years from 1996 [79]. Within 6 h, the US National Ice Center routinely receives ScanSAR images from the Alaska SAR Facility, the Gatineau and Tromsø Satellite Station almost, which provides total Arctic coverage [18]. The sea ice analysis is made by integrating all available remote sensing and *in situ* data, using the SUN SPARC and Ultra Workstations, and a system of satellite image processing. The RADARSAT improved the Ice Patrol's reconnaissance efficiency, although the radar iceberg identification remains problematic even with modern techniques. The RADARSAT ScanSAR wide data provide a daily coverage of the Canadian Arctic, and higher resolution modes are used for sea ice monitoring near the ports, in several selected routes and in the rivers. SAR images are synthesised at the receiving stations Prince Albert and Gatineau and are transmitted to the Ice Centre within 2.5 h to be processed and transmitted to the icebreakers of the Canadian Coast Guard and the department of ice operations for visualisation and analysis. Sea ice monitoring is the most successful online application of the RADARSAT data in Canada, which provides the best combination of geographic coverage and resolution to save about 6 million dollars annually, as compared with airborne radar survey [38]. From February 1996 until the end of 2003, CIS used approximately 25,000 scenes for this purpose [42]. During 2003, a special service carried out iceberg detection and monitoring from satellite SAR imagery, and the International Ice Patrol was the user of this information [42]. Now the RADARSAT ScanSAR imagery is the main data source for sea ice mapping in the Greenland waters. Wind conditions may be an important limitation to the operational use of radar satellite imagery in this area. Small (<50 m across) yet thick ice in concentrations less than 7/10 are frequently undetectable on radar images as they are obscured by a strong backscatter from the sea waves. Therefore, active research into filtering and enhancement techniques has been undertaken to improve discrimination between ice and water [48,49].

The ENVISAT ASAR imagery with almost the same swath as that of the RADARSAT ScanSAR in the VV- and HH-polarisations is an example of further development of SAR technology. The wide swath mode of the ENVISAT satellite is especially suitable for sea ice monitoring, providing a practically daily coverage of most of the Arctic with a high spatial resolution. In mid-2003, the Canadian Ice service began to receive the ENVISAT ASAR data to be used as an additional source to the RADARSAT-1 data for routine production of ice charts, bulletins and forecasts [43].

The Nansen Centres in Bergen and St Petersburg, in collaboration with the European Space Agency and Murmansk Shipping Company, have done a series of projects to demonstrate the possibilities of SAR data for sea ice monitoring and for supporting navigation in the Northern Sea Route (NSR) [64–66]. The NSR, which is a major Russian transport corridor in the Arctic, includes routes suitable for ice navigation confined to the entries to the Novaya Zemlya straits and to the meridian north of Cape Zhelaniya in the west and to the region of the Bering Strait in the east. In August 1991, just after the launch of the ERS-1 satellite, SAR imagery was transmitted in near-real time aboard the French vessel L'Astrolabe via the INMARSAT communication system during her voyage from Europe to Japan in selecting her route in ice [66]. During the period from July 1993 to September 1994, the European Space Agency provided approximately 1000 SAR scenes for sea ice monitoring. Three specific demonstration campaigns in the NSR in the periods of freeze-up, winter and late summer, revealed the ERS SAR capability to map the key ice parameters. The SAR imagery was successfully used to solve tasks of navigation through hard ice. In 1996 the ESA and the Russian Space Agency initiated their first joint project, named ICEWATCH with an overall objective to integrate SAR data into the Russian sea ice monitoring system to support ice navigation in the NSR [65]. During January–February 1996, an experiment was made aboard the icebreakers Vaygach and Taymyr, when the ERS-1 and ERS-2 SARs were operating in a 'Tandem mission', giving a unique opportunity to have SAR coverage over the same area with only a 1-day interval. However, the narrow 100 km swath of the ERS SAR resulted in a substantial spatial and temporal discontinuity in coverage [64].

In August–September 1997, the RADARSAT ScanSAR data were used to support the icebreaker Sovetsky Soyuz operations in the Laptev Sea [119]. With its wide swath, the ScanSAR provided a much better coverage than the ERS SAR, and the selection of scenes along a given ship route was simplified significantly. The ScanSAR data proved to be a very useful supplement to conventional ice maps and could contribute significantly to the ice information. Starting from April 1998, the ScanSAR and the ERS-2 SAR data were acquired and analysed to support the expeditions aboard the icebreaker Sovetsky Soyuz from Murmansk to the Yenisey Gulf [4] and the EC ARCDEV expedition with the Finnish tanker *Uikku* and the icebreaker *Kapitan Dranitsyn* from Murmansk to Sabeta in the Ob River [107]. Throughout the expedition, ScanSAR imagery, aboard the icebreaker was used to detect some important ice parameters, such as the ice types, old and fast ice boundaries, flaw polynyas, wide leads, single ice floes and large areas of rough ice and to solve tactical tasks of navigation. Areas of level and deformed fast ice were identified in the Ob estuary, and an optimal sailing route was selected through the areas with level ice [107]. These expeditions clearly showed that ScanSAR imagery is particularly important for supporting navigation in difficult ice conditions, such as those in the Kara Sea during April–May 1998.

During the summer of 2003, the ENVISAT Wide Swath ASAR imagery was acquired and transmitted aboard the icebreaker Sovetsky Soyuz during her voyage in the Kara Sea, together with visible AVHRR NOAA images. The satellite images and ice maps were displayed in the electronic cartographic navigation system, such that the navigator could see the current icebreaker location overlaid on a satellite image and ice chart in order to select the sailing route.

A series of demonstration campaigns conducted in the NSR since 1991 have shown that high-resolution light- and weather-independent SAR imagery can be effectively used for sea ice monitoring. The sea ice conditions were interpreted and found quite useful for selecting a sailing route. The speed of convoys significantly depends on the ice conditions and varies from about 11–14 knots in polynyas to 4–6 knots in areas with a medium and thick level FY ice and 2 knots in heavily ridged ice [4]. The onboard use of satellite SAR imagery significantly increases the convoy speed in the pack ice (Fig. 9.1). High-latitude telecommunication systems are the main 'bottleneck' in using SAR imagery aboard the icebreakers operating in the NSR. It must be averaged and compressed to about 100–200 kB for their digital transmission. During the first half of 2004, the ENVISAT ASAR imagery was used for sea ice monitoring of the NSR on an experimental basis. Preliminarily processed images were transferred by e-mail to the Murmansk Shipping Company and then were transmitted via the TV channels of the Orbita system to the nuclear icebreakers Yamal, Sovetsky Soyuz, Arktika, Vaygach and Taymyr. The icebreaker navigators could interpret

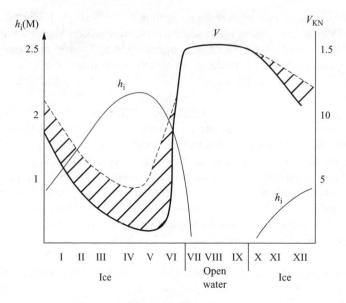

Figure 9.1 *The mean monthly convoy speed in the NSR changes from V_0 (without satellite data) to V_1 (SAR images used by the icebreaker's crew to select the route in sea ice). The mean ice thickness (h_i) is shown as a function of the season. (N. Babich, personal communications)*

them, adequately selecting the easiest sailing through level thin ice and along leads and polynyas with prevailing nilas and grey ice. As a result, the speed of convoys' steering increased by 40–60 per cent on average.

9.1.2.2 Interpretation of satellite SAR imagery of sea ice

A successful application of SAR imagery to support navigation required the ability to recognise the major sea ice parameters and processes from them. Characteristic signatures of major sea ice types and features in ERS, RADARSAT and ENVISAT SAR imagery were described and validated with subsatellite data during field campaigns.

The major stages of ice development described in the WMO Ice Nomenclature include new ice, nilas, young, first-year and old ice. The sea ice recently formed on the water surface may have dark and light SAR signatures. The grease ice that represents an agglomeration of frazil crystals into a soupy layer precludes the formation of short waves (Fig. 9.2(a)) and can be detected as dark stripes and spots among bright SAR signatures of wind-roughened water surface (Fig. 9.2(b)). Slush and shuga have a high backscatter coefficient due to their rough surface and are seen in the SAR images as bright elongated stripes. Nilas represents an elastic ice crust less than 10 cm thick, bounding under the wave action (Fig. 9.3); it has a low backscatter coefficient and a dark SAR signature (Fig. 9.4). Young ice represents the next stage of development; it is subdivided into grey ice and grey–white ice with thicknesses of 10–15 cm and

Figure 9.2 (a) Photo of grease ice and (b) a characteristic dark SAR signature of grease ice. ©*European Space Agency*

200 *Radar imaging and holography*

Figure 9.3 Photo of typical nilas with finger-rafting

15–30 cm, respectively. During winter, young ice is quite common in polynyas and fractures. It has a relatively high backscatter coefficient [102] and can be distinguished from both nilas and first-year ice due to its bright SAR signature (Fig. 9.4). The first-year ice, which is subdivided into thin (30–70 cm), medium (70–120 cm) and thick (over 120 cm) first-year ice, has a typical dark tone. It is difficult to separate thin, medium and thick first-year ice using only their SAR signatures, so knowledge of sea ice conditions in different Arctic regions is used to partly solve this problem. Old ice that has survived melting during at least one summer, is often reliably discriminated from first-year ice due to its brighter tone, rounded floes and distinctive texture (Fig. 9.5). When old and first-year ice breaks into small ice floes with size less than the SAR spatial resolution, their separation is impossible. SAR signatures of second-year and multiyear ice are quite similar, and it is hard to distinguish these types of ice [102].

The backscatter from the ice of the same age depends on its prevailing forms (floe size) and surface roughness. Pancake ice has a rough surface due to characteristic raised pancake rims at the plate edges that lead to a high backscatter and a bright tone in a SAR image (Fig. 9.6). Areas of small ice floes unresolved by radar may have a specific bright SAR signature. When the size of ice floes greatly exceeds the radar spatial resolution, they can be detected in SAR imagery. Single ice floes of even relatively small size can be detected from the dark tone on a bright radar image of wind-roughened water surface, whereas their detection in calm water surface is more difficult. The analysis of ice floes becomes complicated when they touch each other [120,128]. The backscatter of deformed ice is much higher than that of level ice,

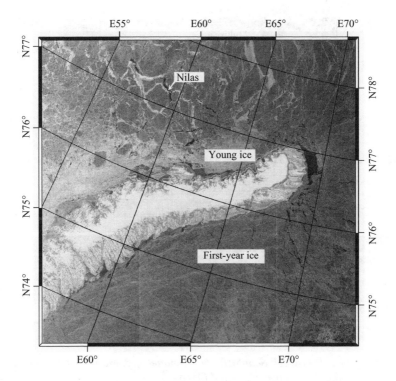

Figure 9.4   A RADARSAT ScanSAR Wide image of 25 April 1998, covering an area of 500 km × 500 km around the northern Novaya Zemlya. A geographical grid and the coastline are superimposed on the image. ©Canadian Space Agency

therefore, areas of weakly, moderately and strongly deformed ice are detectable in ERS, RADARSAT and ENVISAT SAR imagery (Fig. 9.7). Identification of strongly deformed ice hazardous to navigation is particularly important.

Detection of open water areas among sea ice, such as fractures, leads and polynyas, is necessary for selection of an icebreaker's route. Shore and flaw polynyas can be detected reliably, and their width, as well as the type of sea ice can be determined. For example, flaw polynya along the western coast of Novaya Zemlya is clearly evident in RADARSAT ScanSAR imagery (Fig. 9.4) together with a number of fractures covered with nilas (dark tone) or young ice (light tone). It was found that the detection of 100-m wide leads in compact first-year ice is feasible in ScanSAR images.

In winter, fast ice covers large areas in the coastal zones of the Eurasian Arctic Seas. The SAR signature of fast ice is similar to that of drifting ice, and it changes with the surface roughness and, to some degree, with salinity. Level fast ice has a uniformly dark tone, and its boundary can often be identified in SAR images (Fig. 9.7).

The ice edge presents a boundary between open water and sea ice of any type and concentration; it may be both compact and diverged, separating open ice from water.

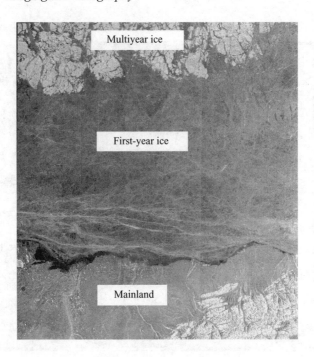

Figure 9.5 A RADARSAT ScanSAR Wide image of 3 March 1998, covering the boundary between old and first-year sea ice in the area to north Alaska. ©Canadian Space Agency

The ice edge may be well-defined or diffuse, straight or meandering, with ice eddies and ice tongues, extending into open water [67]. Ice tongues at the ice edge in the Barents Sea are evident in ENVISAT ASAR imagery (Fig. 9.8). With frequent SAR images, one can investigate the ice edge development in much detail [120]. The sea ice concentration and ice edge location are the most important parameters during the summer; they can be derived from SAR images together with large ice floes, stripes of ice in water, ice drift vectors and areas of convergence/divergence [119].

A high-resolution SAR is considered to be an optimal remote sensing instrument for detection of icebergs. Its backscatter coefficient significantly exceeds that of sea ice and calm sea surface; icebergs that are much larger than the radar spatial resolution are evident as bright spots. In some cases, iceberg shadows and tracks in the sea ice can be detected [125]. Identification of smaller icebergs is complicated by speckle-noise of SAR systems. Areas of iceberg spreading in Franz Josef Land, east of Severnaya Zemlya, and in the northwest Novaya Zemlya have been identified from ERS and RADARSAT SAR data. ERS-2 SAR imagery of Severnaya Zemlya (Fig. 9.9) shows a number of icebergs as bright spots in the Red Army Strait.

Recent studies have shown that the sea ice classification can be improved by using the ENVISAT alternating polarisation mode. Cross-polarisation will improve

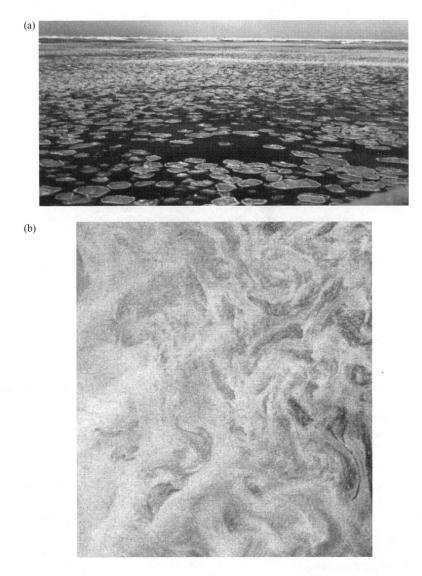

Figure 9.6 *(a) Photo of a typical pancake ice edge and (b) a characteristic ERS SAR signature of pancake ice. A mixed bright and dark backscatter signature is typical for pancake and grease ice found at the ice edge. ©European Space Agency*

the potential for distinguishing ice from open water, which can sometimes be difficult to do only with HH or VV polarisation. In addition to the backscatter variation in single polarisation data, a proper combination of VV and HH dual polarisation and cross-polarisation imagery provides additional information on the sea ice parameters [54,101,122].

204 *Radar imaging and holography*

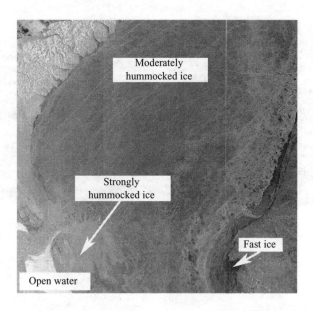

Figure 9.7   A RADARSAT ScanSAR Wide image of 8 May 1998, covering the south-western Kara Sea. ©Canadian Space Agency

Some of the sea ice parameters cannot be found from SAR imagery. For example, it is quite difficult to distinguish thin, medium and thick first-year ice, or second-year and multiyear ice types. It is impossible to determine the snow depth on sea ice and some other parameters. In some cases large ridges and narrow leads covered with grey ice may have similar SAR signatures.

9.1.2.3   Conclusions

The studies have clearly shown that a satellite SAR is a powerful instrument for sea ice monitoring, and SAR data are widely used for this purpose in countries with a perennial or seasonal ice cover. Modern SARs provide a practically daily coverage of the Arctic regions. The most important sea ice parameters can be derived from SAR imagery, and their use increases the safety of navigation and speeds of convoys in severe Arctic ice conditions.

9.1.3   *SAR imaging of mesoscale ocean phenomena*

The SAR imagery allows a global view of most oceanographic phenomena: waves, currents, fronts, eddies and slicks reveal hidden features (such as internal wave and bottom topography). Although most of the imaging mechanisms are now well understood, there are still gaps in our knowledge of certain details. Some aspects still remain obscure, requiring further research efforts.

A high spatial resolution and sensitivity of modern satellite SAR systems makes it possible to observe mesoscale and small-scale features of the sea surface. This allows the use of SAR imagery for investigation of wind speed over the open ocean and

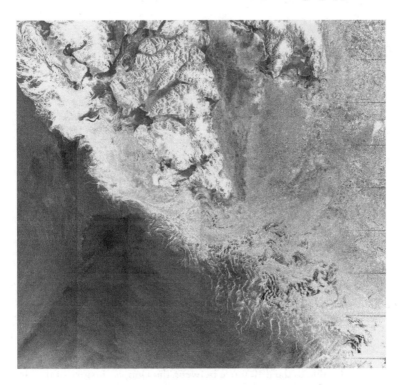

Figure 9.8 An ENVISAT ASAR image of 28 March 2003, covering the ice edge in the Barents Sea westward and southward of Svalbard. ©European Space Agency

coastal zone, surface roughness characteristics and surface polluted zones of different nature. SAR data help to monitor ocean dynamic processes, frontal boundaries, convergence zones, etc.

The normalised radar cross-section (NRCS) is a measure of intensity of the echo signal. In the range of the microwave frequencies, a radar is sensitive to small perturbations of the ocean surface. The NRCS is directly related to the sea roughness, that is, to statistical properties of the sea surface. This allows a radar to detect a larger number of near-surface phenomena than any other remote sensing tool. On the other hand, this makes the radar data extremely hard to interpret, especially quantitatively, and requires the use of sophisticated models.

When dealing with the ocean, one has to consider surface velocities. The motion associated with travelling waves affects significantly the SAR imaging mechanisms. In particular, an azimuthal image shift is due to the motion of the target in the range direction. This motion has little effect on the radial velocities and is unaffected by the pulse compression. It is intense enough to have an influence on the aperture. The azimuthal shift and reduction in the signal amplitude are associated with the motion of the target in the range direction. Wave motion in the azimuthal direction is also a source of image degradation but is of less importance. It is known as azimuth

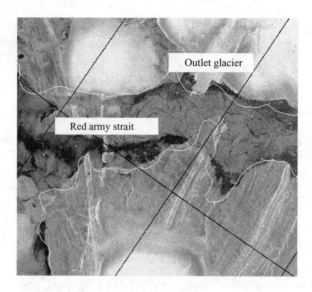

Figure 9.9 An ERS-2 SAR image of 11 September 2001, covering the Red Army Strait in the Severnaya Zemlya Archipelago. ©*European Space Agency*

defocusing and is due to the difference between the Doppler history of the target and the reference signal.

A satellite-borne SAR can monitor large- and small-scale structural fluctuations through the description of the energy distribution of the ocean waves in the spectral domain. The latter is formally described by the wave action balance equation for the spectrum evolution under the combined influence of wind forcing, dissipation, resonant wave–wave interaction, the presence of surfactants and surface current velocity gradients. The possibility of identifying oceanic processes is directly related to changes in the surface scattering characteristics which depend on these processes. For this reason, the detection becomes impossible when no wind is present.

When these phenomena are known, an imaging model can be used to derive the wave spectrum from the image spectrum. Unfortunately, the mechanisms responsible for the spectrum modulation are not fully understood. The analysis of a SAR image is always complicated by interpretation ambiguity. The reason is that one and the same NRCS contrast may be caused by the variation in different physical parameters. Moreover, one and the same phenomenon may manifest itself in some observation conditions and not in others. One of the generally recognised features of radar imagery is the fact that surface phenomena are more clearly observed in the horizontal polarisation than in the vertical one.

A simultaneous study of synchronous SAR images and other data sources (e.g. infrared and visible images, weather maps) helps in getting a correct interpretation. It should be added that since the influence of current velocity gradients, sea surface temperature, surfactant concentration and other environmental parameters on the wind wave spectrum depends upon the wavelength, a radar using a combination of different wavelengths may be quite useful in revealing the mechanisms responsible for the NRCS contrast.

A number of mechanisms have been suggested which are responsible for manifestation of dynamic ocean phenomena in radar images. It is assumed that the wave–current interaction reveals most processes having the scale of the current non-uniformity of about 0.1–10 km. The following phenomena fall into this category: internal waves, current boundaries, convergence zones, eddies and deep-sea convection. The degree of the ocean front manifestation in a SAR is strongly determined by the atmospheric boundary and by its transformation over the sea surface temperature non-uniformities. In any case, the comparative significance of a mechanism depends on the whole set of factors, including the observed process, wind conditions, regional specificity and unknown circumstances (e.g. Reference 16).

Below we give several examples of how different ocean phenomena may become apparent in SAR images. The ERS-2 SAR image in Fig. 9.10, taken on 24 June 2000 over the Black Sea (east of the Crimea peninsula), illustrates the manifestation of temperature fronts, zones of upwelling and slicks of natural films. The fronts are clear from both the bright and dark departures from the background NRCS. As was mentioned before, a correct image interpretation needs additional information. Figure 9.11 shows the sea surface temperature (SST) from the NOAA AVHRR data a few hours after ERS-2 passage. It gives the temperature distribution helpful in image interpretation. The spatial resolution of the infrared image is 1 km as compared with 100 m provided by a SAR. An upwelling is clearly visible in the upper right corner black partially covered with clouds (with SST about 16°C). The black square is the position of the SAR image and the black curved lines are the distinctive features taken from the SAR image. There appears to be a remarkable correlation between the features in the SST and NRCS fields. The insignificant shift is due to the difference in the time of imaging.

The dark region in the upper left corner of the SAR image shows upwelling, when strong winds force the warm water of the upper layer away from the shore and the cold deep water comes up from below. Upwellings are known to occur quite often near the region of the Crimean shoreline. A patch of cold water manifests itself through a modulation of the so-called friction velocity. This quantity may be described as 'effective wind' because it is friction velocity determining the energy flux from the wind to the waves. The stratification of the atmospheric boundary layer over cold water is more stable than over the surrounding warm water. This results in a lower friction velocity, which means that the wind of the same speed (at a given height) would generate less waves in cold water than in warm water. Surface roughness of the upwelling zone is decreased reducing its NRCS. Other conditions being equal, cold water will appear darker than warm water on a radar image (e.g. Reference 16). This feature allows a radar to sense the temperature non-uniformities of the sea surface in general.

There are dark stretched features all over the SAR image. The accumulation of surfactants is assumed to be the cause of these areas of low backscatter. It may take place in regions of high biological activity. When natural (organic) substances reach

208 *Radar imaging and holography*

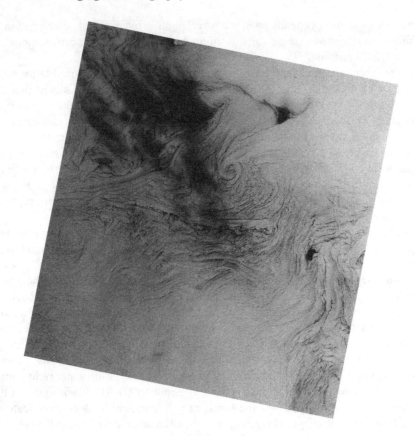

Figure 9.10 *An ERS-2 SAR image (100 km × 100 km) taken on 24 June 2000 over the Black Sea (region to the East Crimea peninsula) and showing upwelling, natural films*

the surface, they tend to be adsorbed at the air–water interface and remain there as a microlayer. Waves travelling across a film-covered surface compress and expand the film, giving rise to surface tension gradients, which lead to vertical velocity gradients within the surface layers. This induces viscous damping and attenuation of short Bragg waves. As a result, the scattered signal returning to the SAR is very much reduced. Natural films are usually dissolved at wind speeds above 7 m/s. Because currents easily redistribute them, such slicks often configure into spatial structures related to the surface current circulation pattern.

Figure 9.12 illustrates how very long ocean waves, the swell, are imaged by a SAR. This image was obtained on 30 September 1995 over the Northern Sea; the land on the right is the Norwegian coast.

We have pointed out that ocean surface roughness of the centimetre scale is due to the local wind (wind stress). Small-scale roughness is modulated by large-scale structures (longer waves or swells). Three mechanisms are considered to be

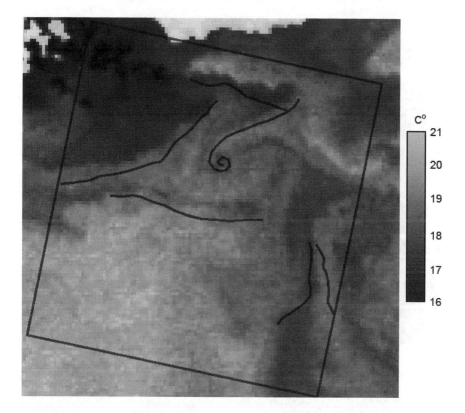

Figure 9.11   SST retrieved from a NOAA AVHRR image on 24 June 2000.

responsible for the longer wave imaging: the tilt modulation, the hydrodynamic effect and velocity bunching. The first mechanism is that long waves tilt the resonant ripples so that the local incident angle changes, modifying the backscatter. The hydrodynamic interaction between the long waves and the scattering ripples lead to the accumulation of scatterers on the up-wind face of the swell. This effect is greatest (as for the tilt modulation) for range travelling waves, and there is no modulation if the ripples are perpendicular to the swell. These first two mechanisms, responsible for swell manifestation, reveal themselves in both synthetic and real aperture imagery. The latter – the so-called velocity bunching effect – is responsible for swell manifestation in the case of long waves travelling close to the azimuthal direction; this effect is observable only in SAR images.

A SAR creates a high-resolution image by recording the phase and amplitude of the electromagnetic radiation reflected by the scatterers and by processing it with a compression filter. The filter is designed to match the phase perfectly for a static target. For the dynamic ocean surface, the motion of each scatterer within the scene distorts the expected phase function with two important implications. First, the linear component of the target motion shifts the azimuth of the imaged location of each

210 *Radar imaging and holography*

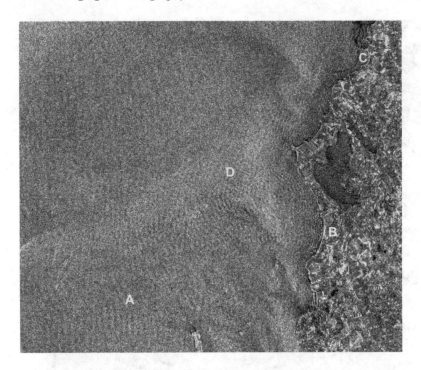

Figure 9.12 *A fragment of an ERS-2 SAR image (26 km × 22 km) taken on 30 September 1995 over the Northern Sea near the Norwegian coast and showing swell*

target. This leads to a strong wave-like modulation in the SAR image due to a periodic forward and backward shift of the scatterer positions. This mechanism is exactly what is known as the velocity bunching. The other implication of the distorted phase function is the degradation of the image azimuthal resolution due to higher order components of the target motion (e.g. Reference 56).

The SAR image enables one to study swell transformation as it approaches the coast. The wavelength decreases as the swell comes to shallow water, so the wavelength is about 350 m at point A while near the coast at point B it is only 90 m (Fig. 9.12). Another observable feature is the swell refraction on the sea bottom relief. This effect is due to the fact that the wave velocity decreases with decreasing depth. The wave crests rotate so as to be parallel to the isobaths. It is clearly visible at points B and C that the swell goes parallel to the curved shore line, though initially it was not. Finally, at point D we can see an interference pattern produced by two swell systems going in approximately perpendicular directions.

Figure 9.13 shows the manifestation of the mentioned ocean features and some new ones. This SAR image was acquired on 28 September 1995 over the Northern Sea.

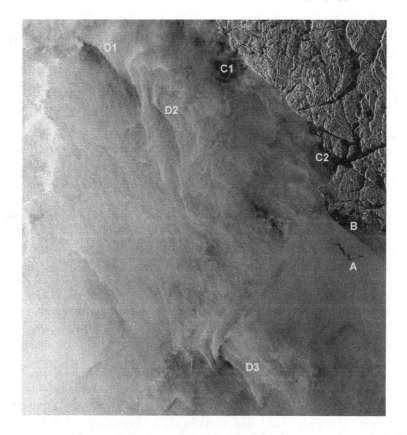

Figure 9.13   *An ERS-2 SAR image (100 km × 100 km) taken on 28 September 1995 over the Northern Sea and showing an oil spill, wind shadow, low wind and ocean fronts*

The first distinctive feature marked as 'A' in Fig 9.13 can definitely be identified as an oil spill. Oil slicks are seen as patches of different shapes with very low NRCS and relatively sharp borders. Quite often, the spill source (ship or oil drill platform) is visible nearby. As compared to natural films, oil films have a higher viscosity, damping short waves more effectively and remaining observable at higher winds when natural slicks would disappear. Another characteristic to distinguish between oil and natural films is that the latter nearly never appear as single localised features but tend to cover vast areas of intricate patterns produced by currents. Anthropogenic oil spills on the sea surface may originate from leaks from ships, offshore oil plants and ship wrecks. In the case of ship wreck, a SAR can contribute to oil spill detection and monitoring, keeping track of the drift and spread of the slicks.

Usually, the shorter the radar wavelength, the more intense is the backscattering reduction due to oil presence. The reduction in the radar backscattering also depends on the incidence angle. Optimum range of angles is defined by the radar wavelength.

One of the strongest obstacles to oil spill detection is the state of the sea. At low (2–3 m/s) wind speeds, SAR images of the ocean become dark because the Bragg scattering waves are not present. In this case almost no features can be distinguished on the sea surface. At high winds, most kinds of oil are dispersed into the water column by the wind waves and also become unobservable (e.g. Reference 39).

The second feature in Fig 9.13 ('B') reveals a clearly lined dark zone near the shore which seems to have the same direction as the dominating wind. The mountainous coastal landscape and the sharp outline allow attributing this feature to wind sheltering by land. It can be seen that the NRCS becomes larger as the distance from the shore along the wind direction increases and the sea roughness becomes better developed. The dark areas 'C1' and 'C2' have blurred contours and may be interpreted as low wind zones.

Besides this, one can see numerous manifestations of the current boundaries ('D1', 'D2', 'D3'). At moderate wind speeds (3–10 m/s), the SAR is capable of revealing the current boundaries, meanders and eddies. The NRCS variation in the vicinity of the current boundary/front is associated with several phenomena, including changes of the stability of the atmospheric boundary layer, wave–current interaction and surfactant accumulation. The exact view of the ocean front on a radar image is affected by many factors: the radar parameters, the observation geometry, the wind conditions, surface current and temperature gradients, etc. Nevertheless, some simple rules of thumb exist. One of them was already mentioned: cold water looks darker than warm water. Another is that convergent current fronts usually appear bright, while divergent fronts appear dark. It is assumed that the features 'D1' and 'D3' are the ocean fronts where the non-uniform current distribution is combined with SST changes. Lack of additional sources of information (e.g. IR images) retains the interpretation ambiguity since a dark area can also be associated with low winds.

Sometimes, atmospheric phenomena may be observable on SAR images, when they affect the near-surface wind. Depending on the observation conditions, such phenomena increase or decrease the radar backscattering by intensifying or damping the Bragg waves. One example is present in the ERS-1 SAR image of Fig. 9.14, taken on 29 September 1995 over the Northern Sea. There are several rain cells of different size scattered throughout the scene. The falling rain drops entrain the air to form a downward flux of cold air. When hitting the ocean surface, the flux transfers cold air mass away from the cell centre to form a wind squall – a line of abrupt increase in the wind speed. The rain cells become visible because the background wind at their boundaries is summed with the wind due to the rain cold air motion. As a result, the wind squall on the lee side of the cell increases the background wind, decreasing it on the opposite side. Thus, one half of the rain cell becomes brighter than the background while the opposite side becomes darker. The distinct boundaries between the wind squalls and the surrounding background water are called squall lines. When the rain is heavy, the centre of a rain cell may appear dark because the falling drops create a turbulence in the upper water layer, damping the Bragg waves. Such phenomena are typical of subtropical regions but may be encountered anywhere else [62].

Figure 9.15 shows a ERS-2 SAR image taken on 30 November 1995 over the Northern Sea. Points 'A', 'B' and 'C' are examples of internal waves on the SAR

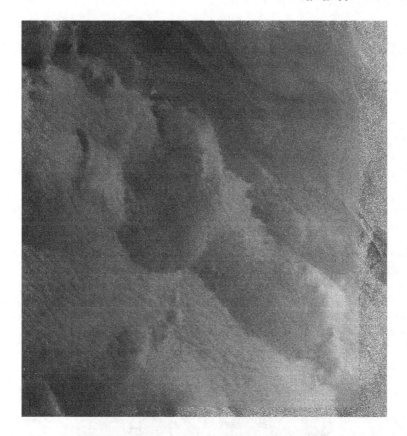

Figure 9.14   An ERS-1 SAR image (100 km × 100 km) taken on 29 September 1995 over the Northern Sea showing rain cells

imagery. Internal waves are one of the most interesting ocean features revealed by SAR imagery. At the beginning of SAR history their detection was entirely unexpected. At present, they are found on SAR images in many regions of the World Ocean at various wind speeds and water depths. They appear as dark crests (troughs) against a lighter background or as light ones against a dark background. The crests always occur as packets called trains. In this image, three trains can be observed. Often, internal waves correlate (parallel) with the bottom topography, when they are caused by the interaction between the tidal currents and abrupt topographic features. The distance between individual dark and light bands varies from several hundred metres to a few kilometres, decreasing from a leading wave to a trailing edge (e.g. [126]).

Orbital motions induced by an internal wave train generate an intermittent pattern of convergent and divergent zones on the sea, which moves with the phase velocity of the internal wave. Convergent zones are generated behind the internal wave crest and divergent zones are behind the troughs. It is these zones that make internal waves visible on radar imagery. There are few commonly accepted explanations about

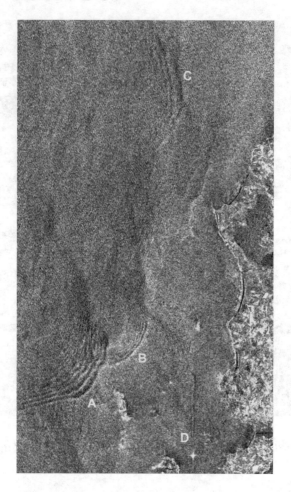

Figure 9.15 An ERS-2 SAR image (18 km × 32 km) taken on 30 September 1995 over the Northern Sea showing an internal wave and a ship wake

how this may happen. According to one point of view, surfactants are accumulated in the convergence zones, which results in short wave damping and makes these zones appear dark on radar images. Another theory states that convergence zones appear bright because these are zones of enhanced roughness due to intensified wave breaking there. The question of which imaging mechanism dominates and under what conditions is still open.

The next distinctive feature clearly observable on the image ('D'), is a ship wake. The ship itself is seen as an extremely bright spot because of many metallic structures that serve as corner reflectors. The wake is a narrow V-shaped feature associated with the ship mark. It appears on radar images only in low wind conditions due to the short lifetime of the Bragg waves and the common ship speeds. The major result of the ship movement is the appearance of the stern wake. This turbulent wake damps the Bragg waves, producing an area of dark return, which is sometimes surrounded by two bright lines. The lines of high backscatter originate from the Bragg waves induced by vortices from the ship's hull. However, there is generally a large diversity of ship wake patterns including combinations of dark and bright stripes on the SAR images and depending on the observational and sea conditions.

Thus, during the last decades the role of SAR data in earth observations has increased considerably, and the SAR has become a major remote sensing tool for environmental monitoring. Improvement of image interpretation techniques, automatised data interpretation, improvement of high-latitude telecommunication systems and a convenient presentation of the information products to the user are necessary for further development of SAR earth monitoring.

9.2 The application of inverse aperture synthesis for radar imaging

The imaging techniques we have discussed in Chapters 5 and 6 did not use holographic or tomographic principles but were developed within a purely radar approach in the United States about 40 years ago. The first device was designed and constructed by the Westinghouse company and represented a narrowband radar with a discrete variation of the carrier frequency and a synthesised spectrum. At about the same time, the Willow Run Laboratory in the United States initiated work on constructing a radar for aircraft imaging; the model radars were tested on an open test ground. Somewhat later, two experimental types of radar were designed for spacecraft identification. One was constructed at the US Air Force Research Center in collaboration with the General Electric Company and the Syracuse Research Corporation (the design of the data processor). The other type of radar was made by the Aerospace Corporation; it had the carrier frequency of 94 GHz, the radiation bandwidth of 1 GHz and the pulse base of 10^6.

The first quality images of low-orbit satellites were obtained by ALCOR radar with the range resolution of 50 cm in the early 1970s. Further efforts by the designers (the Lincoln Laboratory, the Massachusetts Institute of Technology and the Syracuse Research Corporation) to improve this system within a global program for space object identification resulted in the creation, in the late 1970s, of a long-range imaging radar (LRIR) [20,52,83] with better characteristics (Table 9.6).

The major advantages of this radar system are a high-frequency stability, a pulse repetition rate higher than the maximum Doppler frequency of an echo signal, and a controlled repetition rate necessary for time discretisation of transmitted and received pulses. Besides, a LRIR system provides imaging of targets on far-off orbits (including geostationary orbits) and having high rotation rates.

The Doppler-range method of echo signal processing for 2D imaging of the Russian orbiting stations Salut-7 and Kosmos-1686 was implemented in a radar with a 1 GHz probing pulse width [91]. A theoretical and experimental investigation of the imaging of stabilised low-orbit satellites was described in Reference 124, using narrowband probing pulses. The processing algorithms were based on holographic principles. The authors believe that current interest in microwave holography is due to the fact that many available radar systems can acquire a new function – 2D imaging

Table 9.6 The LRIR characteristics

Antenna type (primary reflector)	Paraboloidal
Aperture shape	Circular
Aperture diameter (m)	36.6
Wavelengths (GHz)	K-band
Narrowband mode (NBM)	5.5–6.5
Wideband mode (WBM)	9.5–10.5
Aperture field distribution	Cosine
Sidelobe level (dB)	−22.4
Polarisation (in transmisson and in reception)	Circular
Frequency band (GHz)	1
Pulse duration (μs)	250
Transmitter pulse power in modes 1 and 2 (MW)	0.5, 0.8
Average power (kW)	200
Secondary processing in WBM	Coherent integration
Signal modulation	Linear frequency type
Range gate (m)	30, 60, 120
(frequency filter band (MHz))	(0.8, 1.6, 3.2)
Pulse repetition rate (Hz)	1600 (determined by range measurement unambiguity)
Pulse compressibility	250,000
Sidelobe level of matched filter (in range) (dB)	32
Interpulse instability (°)	3–2
Impulse filling (τ_{imp}/T_{rep}) (%)	50
Way of target tracking	Single pulse
Reception loss (dB)	7.9
Aim of the mode	
NBM	Detection, tracking, range measurement
WBM-1	Target classification
WBM-2	Target classification (from images)
Frequency band (GHz)	Up to 40
Possible radar frequency extension (GHz)	Up to 40
Radar location	Westford, USA (Lincoln Laboratory, space survey facilities)

of space targets – without being radically modernised. An echo signal in such radars is processed by inverse synthesis of microwave holograms owing to the target angle variation during the satellite motion along its orbit. The algorithm uses an original technique for synthesising a 2D image, in the view-flight path plane, from 1D images obtained along a lengthy target path. The summation of partial 1D images produces intensity maxima at the beam interception points corresponding to various angles of the target scatterers. A numerical simulation has shown that this algorithm provides a resolution of about 10 cm for the viewing time of about 2 min.

The experiments on testing this type of radar used a radar interferometer consisting of three antennas of 2.5 m in diameter with a base of 500 m [124]. The antennas were co-phased to provide a coherent transmission and reception of quasi-monochromatic signals with a 4 cm wavelength. The radiation power was 75 kW. The experiment included several observation runs of the Progress spacecraft during its departure from the Mir orbiting station. An optimal 2D image was obtained from 55 1D partial images, each having the synthesis time of about 2 s. The time step between consecutive images was 1.1 s, during which the vision line was rotated by about 0.01 rad. It appeared that some of the scatterers of this nearly cylindrical target were not resolved well enough, the boundaries between them were smeared, and the resulting image represented a bright surface. Still, the image allowed evaluation of the target's dimensions consistent with the real ones.

Therefore, the available radars designed for entirely different applications can be successfully used for spacecraft imaging. For example, the image reconstruction algorithms can operate on the base of a phasometric device originally designed for coordinate measurements. It is important to emphasise that inverse aperture synthesis is also employed successfully in radar viewing of planets. In particular, a pioneering experimental imaging of Venus was carried out by the specialists at the Jet Propulsion Laboratory, California Institute of Technology, USA.

9.3 Measurement of target characteristics

Problems involving the analysis of radar performance require *a priori* information about the scattering properties of a target. These properties are described by a whole combination of independent radar responses to the target of interest. Today, experimental and theoretical investigations of responses is a rapidly developing area of radar science and technology. It involves the search for new forms of description of radiation scattering by various targets and novel methods of their measurement [11,12,30,90,138].

The key position among the many radar responses is occupied by the scattering matrix, which characterises the transformation of the amplitude, phase and polarisation of an arbitrary planar monochromatic wave scattered by a small-size (point) object. The knowledge of the scattering matrix is important for the computation of dynamic and static responses for many applications: the justification of the radar design, the development of methods and devices for antiradar measures, designing of processing algorithms, etc. Besides, a scattering matrix is necessary to go over to responses which describe the target's scattering of probing pulses having complex spectra [138]. It is also indispensable in the computation of local responses to find the scattering properties of individual parts of a target [12].

Theoretically, the exact values of matrix elements can be found only for targets of simple geometry (spheres, cylinders, etc.). So a common way of determining radar responses is by measuring the physical characteristics. For small-size targets, such measurements are commonly made during flight and ground tests. Natural flight tests provide the most complete and reliable data on the target in question but they are very costly and need special equipment and testing conditions.

Radar responses are often measured in special setups on open and closed test grounds. Open tests are carried out either with real targets or their models of natural size. This allows a detailed study of the scattering characteristics and their behaviour under different conditions. However, the response data are often affected by the current weather conditions, background signals from the surrounding objects, natural and artificial noise, etc. Common limitations of an open test ground are the lack of an exact frame of reference for the angular position of the target under study, poor coupling between normally polarised measurement channels, as well as a low data accuracy because of the background effects. Moreover, a measurement run for one target takes a long time, from 4 to 6 h, and is quite costly because of the necessity to maintain the test equipment and facilities.

Closed tests are made in an anechoic chamber (AEC), whose inner walls are covered with a microwave-absorbing material, allowing simulation of wave propagation in free space [98]. But two conditions are to be met in such experiments: the probing wave front is to be planar near the target and the background noise is to be kept below a permissible level. The measurements made in an AEC do not have the limitations of an open ground and take 4–5 times shorter time for one run. Such chambers have found a wide application because they are screened from outside noise, providing an electromagnetic compatibility. Since the electromagnetic, mechanical and climatic conditions in an AEC can be kept constant for a long time, the measurements can be readily automatised and the targets used may be both real objects and models (of natural size or diminished). The choice of the type of target is primarily determined by the size ratio of the target and the so-called echo-free zone in the chamber, that is, the zone where the incident field meets certain requirements as to the wave front geometry and the background signal intensity. This ratio largely determines the response data accuracy. The echo-free zone size is, in turn, determined by the chamber dimensions and the way the wave front is collimated. When the target of choice is larger than the echo-free zone, one usually employs a scaling method, using a smaller model object and a shorter radiation wavelength. One serious disadvantage of this technology is the difficulty of measuring radar responses to targets with absorbing or semiconducting coatings and, sometimes, of making suitable model targets.

The measuring facilities using AECs have some common disadvantages:

1. The measurement accuracy is quite low because of a strong background signal in the chamber working area, associated with the microwave-absorbing materials of high reflectivity (-20 to -30 dB).
2. The echo-free zone is small because the collimators have a small aperture and the chambers a small size; as a result, such measurements cannot be made with real targets.
3. The frequency band of transmitted pulses is limited and bistatic measurements are restricted.

It is clear from this analysis that a closed test ground is preferable for making response measurements for various targets, especially for aircraft and spacecraft. These facilities employ large AECs providing a high accuracy of all matrix elements for a real target, and there is no need to use scaling.

On the other hand, many applied radar problems, especially the estimation of efficiencies of methods and devices for target detection and recognition, often require a numerical simulation of the whole radar channel, including the microwave path, tracking conditions and so on. To do this, one should combine analogue and digital simulation means, including a radar measurement ground (the analogue component) and a computer with appropriate software packages (the digital component). If such equipment is designed for the measurement of reflected signals with their amplitudes and phases, it essentially represents a radar capable of microwave hologram recording, in other words, of inverse aperture synthesis. For imaging, it is sufficient to include in the software the image reconstruction algorithms described in this book.

The next procedure at the imaging stage is the measurement of local responses, or scattering matrices and their elements, to obtain data on individual target scatterers [12,138]. Objects of simple geometry, whose local responses can be calculated precisely, can be used as standards for calibration of measuring devices. Practically, it is reasonable to use cylinders as standard targets. An illustration of the calculation of local responses for cylinders by the EWM suggested by P. Ufimtzev is given in Chapter 2.

The typical measurement facilities include:

- an AEC;
- devices for pulse generation and transmission and for reception of echo signals of various frequencies, including superwideband pulses;
- equipment for making measurements, such as a rotating support, a target rotation control device, etc.
- hard- and software to control measurement runs, to keep records of the incoming and operational data, processors, etc.

The body of work on the measurement of scattering parameters of targets consists of five stages:

- preparatory operations
- preliminary measurements
- major measurements
- control measurements
- data processing.

The preparatory stage is aimed at preparing the measuring devices for a successful performance. Preliminary measurements are to provide information on the device ability to make the necessary measurements, to choose the appropriate operation mode and to calibrate the devices. The aim of the major measurements is to produce microwave holograms of the target with a prescribed accuracy. Control measurements are made in order to check the validity of the data obtained. If the amplitude and phase errors fit into the admissible limits for this particular run, the major measurements are considered to be valid and are fed into a processor together with the calibration data.

Primary processing is performed to bring relative data to their absolute values, that is, to calibrate the measurements and to evaluate the errors. The final results are set into a local database for classified storage. Further processing can be made by various algorithms for the reconstruction of images of different dimensionalities (by using holographic and tomographic processing of the scattering matrix elements) in order to analyse and measure the local responses.

However, the analogue–digital software can also be used for the following tasks:

- to process the results of measurement in order to get statistical data on the scattering characteristics of the target (average values, dispersion, integral distributions, histograms and so on) for given target angles;
- to compute the angular positions of the target during its motion with respect to the ground radar in order to simulate the dynamic behaviour of the echo signal and the radar viewing devices;
- to simulate the target recognition devices by using various methods to find the target recognition parameters (from images, too) and to design decision-making schemes.

As a result, one can get online information about various probable characteristics necessary for the target detection and recognition.

Methods for direct imaging and for measurement of local responses in an AEC are described in detail in Reference 138. So we shall restrict ourselves to a brief review of the measurement procedures and some of the results obtained.

The best way of producing an image in an AEC is to record multiplicative Fourier holograms and to subject them to a digital processing. The recording can be based on one of the schemes shown in Fig. 2.4, and the reconstruction can be made by the algorithm presented in Fig. 9.16.

The input data are two quadrature components $h_{r1}(\varphi)$ and $h_{r2}(\varphi)$ of a 1D complex microwave hologram $h_r(\varphi)$ and the calibration results (the calibration curve). The sampling step for the functions $h_{r1}(\varphi)$ and $h_{r2}(\phi)$ should meet the condition $\Delta\varphi \leq \lambda/l_{max}$, where l_{max} is the maximum linear size of the target.

We can synchronise the quadrature components by using the subroutine for justifying the data file. Normally, a microwave hologram is recorded when the target is rotated by 2 rad and further processing is performed for a sequence of samples, whose number corresponds to the optimal size of the synthetic aperture and the position in the data file corresponds to the required target aspect.

The chosen sequence is normalised, because a microwave hologram can be measured with different receiving channel gain, depending on the recorded signal value. This should be taken into account when measuring a local response in the RCS units. In order to visualise the scatterers and to measure their relative intensities at a given aspect angle, we should reduce the domains of the functions $h_{r1}(\varphi)$ and $h_{r2}(\varphi)$ to $[-1,1]$.

For a direct image reconstruction, one is to use a fast Fourier transform (FFT), which is simple to make when the number of initial readouts is $2m$, where m is

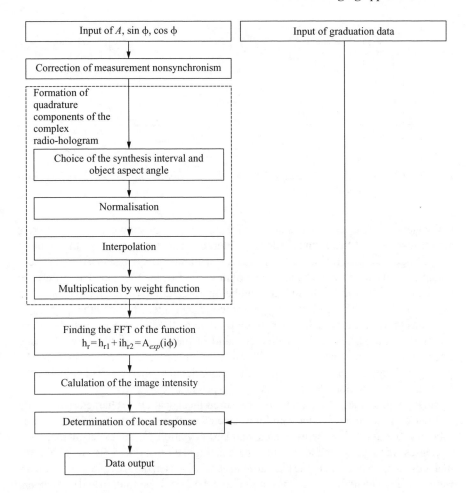

Figure 9.16   The scheme of the reconstruction algorithm

a natural number. Their necessary number is made up of an arbitrary set of initial samples, using an interpolation block. In order to minimise a measurement error in the local response, the chosen sample is multiplied by any weighting function.

Having found the Fourier transform of the complex function $h_r(\varphi)$, we form the files $\text{Re}V$ and $\text{Im}V$, defining the complex amplitudes of the field $V(\nu)$ scattered by the target surface. The image intensity, W, is found as the squared modulus of the function $V(\nu)$. The image sample interval is $\Delta \nu = \lambda/2\psi_s$, where ψ_s is the synthetic aperture angle. With the calibration data, the image intensities of individual scatterers can be represented in the RCS units.

Figure 9.17 illustrates typical 1D images of a perfectly conducting cylinder, obtained in an AEC at the aspect angle $\phi_{\text{obs}} = 105°$, with the image intensity plotted along the ordinate and the normalised target size along the abscissa. The image intensity peaks correspond to the projections of scatterers 1, 2 and 3 onto the normal

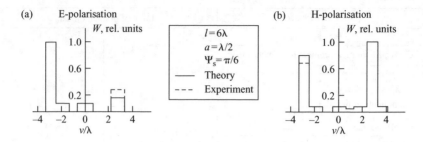

Figure 9.17 A typical 1D image of a perfectly conducting cylinder (l-length of cylinder, a-radius of cylinder)

to the view line (Fig. 2.1). The analysis of these images has shown that the scatterers are localised just at the cylinder edges. Scatterers 2 and 3 at the ends of the cylinder generating line are well resolved. The images of 1 and 2 merge because they are separated by a distance smaller than the resolution limit of the method. The difference in the intensities of individual points can be interpreted in terms of the EWM or the GTD. The dashed lines in Fig. 9.17 are for the former intensities and the latter computations yield similar results. Our findings agree well with experimental data. The polarisation properties of the scatterers manifest themselves in the varying image intensity due to the changes in the illumination polarisation. Such images can be used to estimate the target size and, with a more detailed analysis, its geometry, the 'brightest' construction elements and surface patches.

Figures 9.18 and 9.19 present the measured local scattering characteristics for a metallic cylinder, the RCS diagram for a selected scatterer, and the simulation results (Sections 5.2 and 5.3). The estimated standard deviation for the experimental local responses was 1.8 dB. In addition to a methodological error of 0.5 dB, the total error includes components due to the background echo signals in the AEC, imperfect polarisation channel insulation, etc. It is obvious that the theory, simulation and experiment gave similar results within the accuracy of the total measurement error. Such measurements provide data on local scattering characteristics of targets of complex geometry. The results presented can be used for calibration of measuring setups.

9.4 Target recognition

Recognition of targets is a very important task in radar science and practice. By recognition we mean the procedure of attributing the object being viewed to a certain class in a prescribed alphabet of target classes, using the radar data obtained. According to the general theory of pattern recognition, radar target recognition should include the following stages:

- compiling a classified alphabet of radar targets to be recognised;
- viewing of targets;
- determination (measurement) of some target responses from the recorded echo signal parameters to compile target descriptions, or patterns;

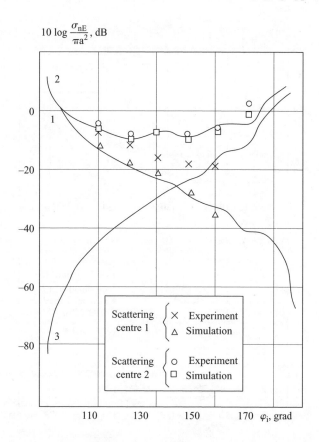

Figure 9.18 The local scattering characteristics for a metallic cylinder (E-polarisation). The subscripts 1, 2, 3 at σ denote scattering centres

- identification and selection of informative signs (features) from the compiled lists;
- target classification or attribution of a particular target to one of the classes on the basis of discriminating signs.

The problem of making up an alphabet of target classes and selecting informative signs to describe each class reliably is quite complicated and is to be solved by qualified and experienced specialists. Of course, classification may be based on various principles. One of them is to group targets in terms of their function and application. For example, a successful management of air traffic needs a classification of aircraft: heavy and light passenger planes, military planes, helicopters, etc.

Each class of radar targets can be described by a definite set of discriminating characteristics to be used for classification: configuration, the presence of well-defined and readily observable parts, dynamic parameters (e.g. altitude, flight velocity), etc. A specific feature of all radar targets is that the radar input senses a target pattern in the echo signal domain. The size scale of this domain and the physical meaning of each of its components differ considerably from those of the parameter vectors of the target

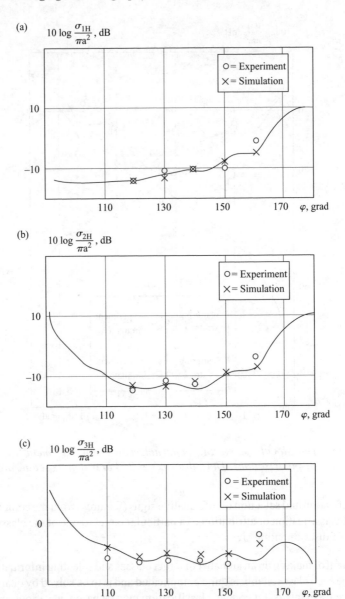

Figure 9.19 The local scattering characteristics for a metallic cylinder (H-polarisation). The subscripts 1, 2, 3 at σ denote scattering centres

class and each characteristic individually. No matter how many identification signs a target possesses, one can get information only about those characteristics that are contained in the recorded echo signal parameters. We believe that a holographic approach to designing target recognition radars is capable of removing this limitation.

The target description (pattern) in a radio vision system is a microwave hologram function, which is generally a vector, non-stationary random function. It is manifested at the radar input as a pattern of a certain class of objects. Such patterns are practically unsuitable for classification because they have a complex probabilistic structure, a large and varying size, etc. Besides, the individual values of the hologram functions may also include minor, unimportant details of a target that may introduce additional recognition errors.

Like in many other target recognition problems, a key task is to reveal the most informative, discriminating target signs. The subsystem of sign identification must include compression and preliminary processing of the initial radar data [12], such that the classification subsystem input would receive a size-fixed array of signs characterising the essential, most typical properties of a particular target. The role of a sign 'identifier' may be played by the operator of image reconstruction from a hologram, which can generally be reduced to an integral Fourier transform. The distances between individual scatterers and the local target characteristics measured from the images will form a discrete vector domain of a relatively small size, whose elements can be considered as recognition signs. They have a clear physical meaning, a factor important for creating a library of standards for the classifier operation. The target recognition then becomes a holographic process with a clear physical meaning. One does not need *a priori* data on the statistical structure of the echo signal, and this method of sign discrimination may be considered as distribution-free.

The final stage in the recognition process is to design a procedure for target classification, that is, finding the criteria for attributing a particular target to a class in a given alphabet. The classification is based on a key rule attributing the array of discriminating signs (i.e. the target itself) to one of the possible target classes. Modern pattern recognition theory has at its disposal a powerful mathematical apparatus including deterministic, probabilistic and heuristic procedures, as well as various sets of criteria for detecting similarities and differences between classes.

Therefore, radar target recognition can be represented as a block diagram that can serve as the basis for a mathematical model of a radar recognition device (Fig. 9.20). This idea has been tested using an analogue and a digital model of a recognition radar. The simulation included the measurement of microwave holograms of different model objects in an AEC, the mathematical modelling of the object motion and the computation of dynamic realisations of the microwave hologram functions with static measurements (for random initial conditions of motion), the modelling of the radar receiving channel, image reconstruction and construction of sign vectors, as well as the classification of the objects.

The relative positions of three 'brightest' scatterers (geometrical characteristics) and their image intensities for each image were found to be

$$R_{12}^{kl} = \left|R_1^{kl} - R_2^{kl}\right|, \quad R_{13}^{kl} = \left|R_1^{kl} - R_3^{kl}\right|, \quad R_{23}^{kl} = \left|R_2^{kl} - R_3^{kl}\right|, \quad A_1^{kl}, A_2^{kl}, A_3^{kl},$$

where $k, l = 1, 2$ are the polarisation indices.

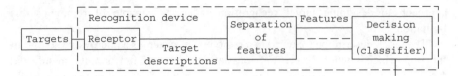

Figure 9.20 A mathematical model of a radar recognition device

The set of sign vectors was stored in the recognition device to be used for creating a teaching or testing standard of sign vectors. The vectors were normalised such that one could compare vectors made up of signs of different physical nature. Smaller-scale sign vectors were created for further use. Table 9.7 presents the vectors for the entire sign domain constructed to minimise the sign vectors and compare their informative characteristics for further recognition. The minimum size was 3 and the maximum 9.

A sequence of recognition sign vectors arrives at the classifier input. We had employed a Bayes classifier and a nonparametric classifier based on the method of potential functions. The former is optimal in the sense that it minimises the average risk of wrong decisions. The teaching of the Bayes classifier included the evaluation of unknown parameters of the conditioned probability distribution of the sign vector x in the class $A_i - p(x/A_i)$, which was taken to be normal. This decision rule is Bayes-optimal at the equal cost of errors for a more general distribution; in practice, however, the difference between the actual and normal distributions is usually neglected if the former is smooth and has one maximum [12]. The other classifier was used when there was no information on the sign vector distribution function. It was assumed that the general decision function was known and its parameters were estimated from the teaching samples [12].

Each experimental run provided a $K \times K$ matrix of decisions (K is the number of classes) at the classifier output. The element k_{ij} of the matrix is the number of objects in the ith class attributed to the jth class. From the matrix K, we can estimate the probability of correct recognition events, the probability of a false alarm, etc.

The model suggested was used to test the recognition capabilities for various objects. We also planned to estimate the efficiency of recognition, to compare the information contents of different sign vectors and investigate the stability of the classification algorithms in terms of the size of the teaching sample. For this, we employed metallic cones with a spherical apex (class 1) and a spherical base (class 2) of about the same length. The probabilistic structure of the sign domain was estimated by constructing experimental holograms. Their unimodal character was tested to justify the use of a Bayes classifier. An experimental series was equal to 100 in all the runs.

Table 9.8 compares the valid recognition probability for objects of both classes and the size of the teaching sequence at different sign vectors for the case of a Bayes classifier. One can see that the largest vectors made up of local responses are most effective. The geometrical characteristics gave poorer results, as was expected, because the objects in both classes were of about the same size. When the number of teaching vectors is decreased, there is a tendency for a lower recognition efficiency.

Table 9.7 The variants of the sign vectors

Type of sign vector	Polarisation				
	1	2	3	4	5
AR	$A_1^{11}, A_2^{11}, R_{12}^{11}$	$A_1^{22}, A_2^{22}, R_{12}^{22}$	$A_1^{12}, A_2^{12}, R_{12}^{12}$	$A_1^{11}, A_2^{11}, R_{12}^{11} A_1^{22}, A_2^{22}, R_{12}^{22}$	$A_1^{11}, A_2^{11}, R_{12}^{11} A_1^{22}, A_2^{22}, R_{12}^{22} A_1^{12}, A_2^{12}, R_{12}^{12}$
A	$A_1^{11}, A_2^{11}, A_3^{11}$	$A_1^{22}, A_2^{22}, A_3^{22}$	$A_1^{12}, A_2^{12}, A_3^{12}$	$A_1^{11}, A_2^{11}, A_3^{11} A_1^{22}, A_2^{22}, A_3^{22}$	$A_1^{11}, A_2^{11}, A_3^{11} A_1^{22}, A_2^{22}, A_3^{22} A_1^{12}, A_2^{12}, A_3^{12}$
R	$R_{12}^{11}, R_{13}^{11}, R_{23}^{11}$	$R_{12}^{22}, R_{13}^{22}, R_{23}^{22}$	$R_{12}^{12}, R_{13}^{12}, R_{23}^{12}$	$R_{12}^{11}, R_{13}^{11}, R_{23}^{11} R_{12}^{22}, R_{13}^{22}, R_{23}^{22}$	$R_{12}^{11}, R_{13}^{11}, R_{23}^{11} R_{12}^{22}, R_{13}^{22}, R_{23}^{22} R_{12}^{12}, R_{13}^{12}, R_{23}^{12}$

Table 9.8   The valid recognition probability (a Bayes classifier)

Number of teaching vectors	Type of sign vector	Polarisation				
		1	2	3	4	5
50	AR	0.54	0.68	0.68	0.68	0.80
	A	0.63	0.61	0.66	0.63	0.78
	R	0.56	0.68	0.55	0.66	0.63
40	AR	0.56	0.64	0.63	0.61	0.77
	A	0.60	0.61	0.67	0.63	0.78
	R	0.56	0.68	0.55	0.66	0.63
30	AR	0.53	0.63	0.52	0.63	0.78
	A	0.60	0.60	0.69	0.64	0.77
	R	0.57	0.67	0.51	0.64	0.63
20	AR	0.53	0.63	0.51	0.64	0.71
	A	0.57	0.58	0.68	0.63	0.73
	R	0.52	0.70	0.52	0.62	0.63
10	AR	0.44	0.59	0.58	0.60	0.50
	A	0.52	0.55	0.59	0.60	0.55
	R	0.50	0.68	0.56	0.54	0.50

Table 9.9   The valid recognition probability (a classifier based on the method of potential functions)

Number of teaching vectors	Type of sign vector	Polarisation				
		1	2	3	4	5
30	AR	0.88	0.90	0.87	0.83	0.81
20		0.82	0.71	0.71	0.78	0.72
10		0.75	0.63	0.67	0.76	0.71
30	A	0.87	0.90	0.90	0.89	0.94
20		0.67	0.82	0.85	0.81	0.84
10		0.64	0.80	0.72	0.75	0.82
30	R	0.80	0.84	0.69	0.72	0.80
20		0.62	0.66	0.54	0.53	0.68
10		0.55	0.60	0.56	0.52	0.63

Table 9.9 shows similar results for a classifier based on the method of potential functions. The recognition efficiency is higher but the time necessary for the teaching is an order of magnitude longer.

The sequence of operations in this model can be used as a procedure for an estimation of recognition efficiency for various targets at the stage of designing the radar or the targets. This model provides a greater efficiency of pre-tests at the device designing stage because one can

- obtain statistical data on possible recognition of various targets in a short time at lower cost;
- get teaching or experimental sequences of practically any size;
- evaluate the effective parameters of antirecognition devices during direct statistical experiments, etc.

References

1 AKHMETYANOV, V. R., and PASMUROV, A. Ya.: 'Radar imaging analysis based on theory of information'. Proceedings of sixth All-Union seminar on Optical information processing, Frunze, USSR, 1986, part 2, p. 59 (in Russian)
2 AKHMETYANOV, V. R., and PASMUROV, A. Ya.: 'Radar imagery processing for earth remote sensing', *Zarubezhnaya Radioelectronica*, 1987, **1**, pp. 70–81 (in Russian)
3 ALEXANDROV, V. Y., LOSHCHILOV, V. S., and PROVORKIN, A. V.: 'Studies of icebergs and sea ice in Antarctic using "Almaz-1" SAR data', in POPOV, I. K., and VOEVODIN, V. A. (Eds): 'Icebergs of the world ocean' (Hydrometeoizdat, St Petersburg, 1996), pp. 30–36 (in Russian)
4 ALEXANDROV, V. Y., SANDVEN, S., JOHANNESSEN, O. M., PETTERSSON, L. H., and DALEN, O.: 'Winter navigation in the Northern Sea Route using RADARSAT data', *Polar Record*, 2000, **36** (199), pp. 333–42
5 ALLAN, T. D. (Ed.): 'Satellite microwave remote sensing' (John Wiley & Sons, New York, 1983)
6 ALPERS, W., and HENNINGS, I.: 'A theory of the imaging mechanisms of underwater bottom topography by real and synthetic aperture radar', *Journal of Geophysical Research*, 1984, **89**, pp. 10529–46
7 ARSENOV, S. M., and PASMUROV, A. Ya.: 'Investigation of local scattering characteristics of lumped objects from their radar images'. Proceedings of All-Union symposium on Waves and diffraction, Moscow, USSR, 1990, pp. 153–55 (in Russian)
8 ARSENOV, S. M., and PASMUROV, A. Ya.: 'Signal processing for aircraft radar imaging', *Zarubezhnaya Radioelectronica*, 1991, **1**, pp. 71–83 (in Russian)
9 ARSENOV, S. M., and PASMUROV, A. Ya.: 'Tomographic signal processing for ISAR', in GUREVICH, S. B. (Ed.): 'Optical and optico-electronic means of data processing' (USSR Academy of Sciences, Leningrad, 1989), pp. 258–66 (in Russian)
10 ARSENOV, S. M., and PASMUROV, A. Ya.: 'Compensation of aircraft radial motion for ISAR'. Proceedings of second All-Union conference on Theory and practice of spatial-time signal processing, Sverdlovsk, USSR, 1989, pp. 217–19 (in Russian)

11 ASTANIN, L. Yu., and KOSTYLEV, A. A.: 'Ultrawideband radar measurements. Analysis and processing' (The Institution of Electrical Engineers, London, 1997)
12 ASTANIN, L. Yu., KOSTYLEV, A. A., ZINOVIEV, Yu. S., and PASMUROV, A. Ya.: 'Radar target characteristics: measurements and applications' (CRC Press, Boca Raton, 1994)
13 AUSHERMAN, D. A., KOZMA, A., WALKER, J. L., JONES, H. M., and POGGIO, E. C.: 'Development in radar imaging', *IEEE Transactions on Aerospace and Electronic Systems*, 1986, **AES-20** (4), pp. 363–99
14 BAKUT, P. A., BOLSHAKOV, I. A., GERASIMOV, B. M. *et al.*: 'Statistical theory of radiolocation' (Sovetskoe radio, Moscow, 1963, vol. 1) (in Russian)
15 BATES, R. H. T., GARDEN, K. L., and PETERS, T. M.: 'Overview of computerized tomography with emphasis on future developments', *Proceedings of IEEE*, 1983, **71** (3), pp. 356–72
16 BEAL, R., KUDRYAVTSEV, V., THOMPSON, D. *et al.*: 'The influence of the marine atmospheric boundary layer on ERS-1 synthetic aperture radar imagery of the Gulf Stream', *Journal of Geophysical Research*, 1997, **102** (C3), pp. 5799–5814
17 BELOCERKOVSKY, S. M., KOCHETKOV, Yu. A., KRASOVSKY, A. L., and NOVITSKIY, V. V.: 'Introduction in aeroautoelasticity' (Nauka, Moscow, 1980) (in Russian)
18 BERTOIA, C., FALKINGHAM, J., and FETTERER, F.: 'Polar SAR data for operational sea ice mapping', in TSATSOULIS, C., and KWOK, R. (Eds): 'Analysis of SAR data of the polar oceans. Recent advances' (Springer-Praxis, Berlin, Heidelberg, 1998), pp. 201–34
19 BORN, M., and WOLF, E.: 'Principles of optics' (Pergamon Press, New York, 1980)
20 BROMAGHIM, D. R., and PERRY, J. P.: 'A wideband liner FM ramp generator for the long-range imagery radar', *IEEE Transactions on Microwave Theory and Techniques*, 1978, **MTT-26** (5), pp. 322–25
21 BROWN, W. M., and FREDERICKS, R. J.: 'Range-Doppler imaging with motion through resolution cells', *IEEE Transactions on Aerospace and Electronic Systems*, 1969, **AES-5** (1), pp. 98–102
22 BROWN, W. M., and GHIGLIA, D. C.: 'Some methods for reducing propagation-induced phase errors in coherent imaging systems', *Journal of the Optical Society of America*, 1988, **5** (6), pp. 924–41
23 BROWN, W. M., and RIORDAN, J. E.: 'Resolution limits with propagation phase errors', *IEEE Transactions on Aerospace and Electronic Systems*, 1970, **AES-6** (5), pp. 657–62
24 BROWN, W. M.: 'Synthetic aperture radar', *IEEE Transactions on Aerospace and Electronic Systems*, 1967, **AES-3** (2), pp. 217–30
25 BUNKIN, B. V., and REUTOV, A. P.: 'Trends of radar development', in SOKOLOV, A. V. (Ed.): 'Problems of perspective radiolocation' (Radiotekhnika, Moscow, 2003), pp. 12–19 (in Russian)

26 BYKOV, V. V.: 'Digital modelling for statistical radio engineering' (Sovetskoe Radio, Moscow, 1971) (in Russian)

27 CARSEY, F., HARFING R., and WALES, C.: 'Alaska SAR facility: The US center for sea ice SAR data', in TSATSOULIS, C., and KWOK, R. (Eds): 'Analysis of SAR data of the polar oceans. Recent advances' (Springer-Praxis, Berlin, Heidelberg, 1998), pp. 189–200

28 CHERNYKH, M. M., and VASILIEV, O. V.: 'Experimental estimation of aircraft echo signal coherence', *Radiotekhnika*, 1999, **2**, pp. 75–78 (in Russian)

29 COLLIER, R. J., BURCKHARDT, C. B., and LIN, L. H.: 'Optical holography' (Academic Press, New York, London, 1971)

30 CURLANDER, I. C., and Mc DONOUGH, R. N.: 'Synthetic aperture radar systems and signal processing' (John Wiley & Sons, New York, London, 1991)

31 CURRIE, N. C. (Ed.): 'Radar reflectivity measurement: techniques and applications' (Artech House, Norwood, USA, 1989)

32 CUTRONA, L. J., LEITH, E. N., PORCELLO, L. J., and VIVIAN, W. E.: 'On the application of coherent optical processing techniques to synthetic aperture radar', *Proceedings of IEEE*, **54** (8), 1966, pp. 1026–32

33 DA SILVA, J. C. B., ROBINSON, I. S., JEANS, D. R. G., and SHERWIN, T.: 'The application of near-real-time ERS-1 SAR data for predicting the location of internal waves at sea', *International Journal of Remote Sensing*, 1997, **18** (10), pp. 3507–17

34 DESAI, M., and JENKINS, W. K.: 'Convolution back – projection image reconstruction for synthetic aperture radar'. Proceedings of IEEE International symposium on Circuits and systems, Montreal, Canada, 1984, vol. 1, pp. 161–63

35 DESCHAMPS, G.: 'About microwave holography', *Proceedings of IEEE*, **55** (4), 1967, pp. 58–59

36 DIKINIS, A. V., IVANOV, A. Y., KARLIN, L. N. *et al.*: 'Atlas of synthetic aperture radar images of the ocean acquired by ALMAZ-1 satellite' (GEOS, Moscow, 1999) (in Russian)

37 DRINKWATER, M. R.: 'Satellite microwave radar observations of Antarctic Sea ice', in TSATSOULIS, C., and KWOK, R. (Eds): 'Analysis of SAR data of the polar oceans. Recent advances' (Springer-Praxis, Berlin, Heidelberg, 1998), pp. 35–68

38 EDEL, H., SHAW, E., FALKINGHAM, J., and BORSTAD, G.: 'The Canadian RADARSAT program', *Backscatter*, 2004, **15** (1), pp. 11–15

39 ERMAKOV, S. A., SALASHIN, S. G., and PANCHENKO, A. R.: 'Film slicks on the sea surface and some mechanisms of their formation', *Dynamics of Atmosphere and Ocean*, 1992, **16** (2), pp. 279–304 (in Russian)

40 ESPEDAL, H. A., and JOHANNESSEN, O. M.: 'Detection of oil spills near offshore installations using synthetic aperture radar (SAR)', *International Journal of Remote Sensing*, 2000, **21** (11), pp. 2141–44

41 ESPEDAL, H. A., JOHANNESSEN, O. M., JOHANNESSEN, J. A. *et al.*: 'COASTWATCH'95: A tandem ERS-1/SAR detection experiment of natural film on the ocean surface', *Journal of Geophysical Research*, 1998, **103** (C11), 24969–82

42 FLETT, D., and VACHON, P. W.: 'Marine applications of SAR in Canada', *Backscatter*, 2004, **15** (1), pp. 16–21
43 FLETT, D., De ABREU, R., and FALKINGHAM, J.: 'Operational experience with ENVISAT ASAR wide swath data at the CIS'. Abstracts of ENVISAT Symposium, Salzburg, Austria, 2004, Abstract No. 363
44 FREIDEY, A. I., CONROY, B. L., HOPPE, D. I., and BRANJI, A. M.: 'Design concepts of a 1-MW CW X-band transmit/receiver system for planetary radar', *IEEE Transactions on Microwave Theory and Techniques*, 1992, **MTT-40** (6), pp. 1047–55
45 FROM PATTERN TO PROCESS: The strategy of the Earth observing system. EOS science steering committee report, vol. 2, NASA, 1988
46 FUREVIK, B. R., JOHANNESSEN, O. M., and SANDVIK, A. D.: 'SAR – retrieved wind in polar regions – comparison with *in situ* data and atmospheric model output', *IEEE Transactions on Geoscience and Remote Sensing*, 2002, **GE-40** (8), pp. 1720–32
47 GHIGLIA, D. C., and BROWN, W. D.: 'Some methods for reducing propagation – induced phase errors in coherent imaging systems. II. Numerical results', *Journal of the Optical Society of America*, 1988, **A5** (6), pp. 942–56
48 GILL, R. S., and VALEUR, H. H.: 'Ice cover discrimination in the Greenland waters using first-order texture parameters of ERS SAR images', *International Journal of Remote Sensing*, 1999, **20** (2), pp. 373–85
49 GILL, R. S., VALEUR, H. H., and NIELSEN, P.: 'Evaluation of the RADARSAT imagery for the operational mapping of sea ice around Greenland'. Proceedings of symposium on Geomatics in the era of RADARSAT, Ottava, Canada, 1997, pp. 230–34
50 GOODMAN, J. W.: 'An introduction to the principles and applications of holography', *Proceedings of IEEE*, 1971, **59** (9), pp. 1292–304
51 GOODMAN, J. W.: 'Introduction to Fourier optics' (McGraw-Hill Book Company, New York, 1968)
52 GOUDEY, K. R., and SCIAMBI, A. F.: 'High power X-band monopulse tracking feed for the Lincoln laboratory long-range imaging radar', *IEEE Transactions on Microwave Theory and Techniques*, 1978, **MTT-26** (5), pp. 326–32
53 GRIFFIN, C. R.: 'Image quality parameters for digital synthetic aperture radar'. Proceedings of symposium on RADAR, 1984, pp. 430–35
54 HAAS, C., DIERKING, W., BUSCHE, T., HOELEMANN, J., and WEGENER, C.: 'Monitoring polynya processes and sea ice production in the Laptev sea', Abstracts of ENVISAT Symposium, Salzburg, Austria, 2004, Abstract No. 137
55 HARGER, R. O.: 'Synthetic aperture radar systems. Theory and design' (Academic Press, New York, 1970)
56 HASSELMANN, K., RANEY, R. K., PLANT, W. J. *et al.*: 'Theory of synthetic aperture radar ocean imaging: A MARSEN view', *Journal of Geophysical Research*, 1985, **90** (10), pp. 4659–86
57 HERMAN, G. T.: Image reconstruction from projections. The fundamentals of computerized tomography' (John Wiley & Sons, New York, 1980)

58 ILYIN, A. L., and PASMUROV, A. Ya.: 'Fluctuated objects and SAR characteristics', *Izvestiya vysshykh uchebnykh zavedeniy – Radioelectronica*, 1989, **32** (2), pp. 65–68 (in Russian)
59 ILYIN, A. L., and PASMUROV, A. Ya.: 'Mapping of partial coherence extended targets by SAR', *Zarubezhnaya Radioelectronica*, 1985, **6**, pp. 3–15 (in Russian)
60 ILYIN, A. L., and PASMUROV, A. Ya.: 'Radar imagery characteristics of fluctuated extended targets', *Radiotekhnika i Electronica*, 1987, **31** (1), pp. 69–76 (in Russian)
61 IVANOV, A. V.: 'On the synthetic aperture radar imaging of ocean surface waves', *IEEE Journal of Oceanic Engineering*, 1982, **OE-7** (2), pp. 96–103
62 JOHANNESSEN, J., DIGRANES, G., ESPEDAL, H., JOHANNESSEN, O. M., and SAMUEL, P.: 'SAR ocean feature catalogue' (ESA Publications Division, ESTEC, Noordwijk, The Netherlands, 1994)
63 JOHANNESSEN, J. A., SHUCHMAN, R. A., JOHANNESSEN, O. M., DAVIDSON, K. L., and LYZENGA, D. R.: 'Synthetic aperture radar imaging of upper ocean circulation features and wind fronts', *Journal of Geophysical Research*, 1991, **96** (9), pp. 10411–22
64 JOHANNESSEN, O. M., SANDVEN, S., PETTERSSON, L. H. *et al.*: 'Near-real time sea ice monitoring in the Northern Sea Route using ERS-1 SAR and DMSP SSM/I microwave data', *Acta Astronautica*, 1996, **38** (4–8), pp. 457–65
65 JOHANNESSEN, O. M., VOLKOV, A. M., BOBYLEV, L. P. *et al.*: 'ICE-WATCH – Real-time sea ice monitoring of the Northern Sea Route using satellite radar (a cooperative earth observation project between the Russian and European Space Agencies)', *Earth Observations and Remote Sensing*, 2000, **16** (2), pp. 257–68
66 JOHANNESSEN, O. M., and SANDVEN, S.: 'ERS-1 SAR ice routing of L'Astrolabe through the Northeast Passage', *Arctic News-Record, Polar Bulletin*, **8** (2), pp. 26–31
67 JOHANNESSEN, O. M., CAMPBELL, W. J., SHUCHMAN, R. *et al.*: 'Microwave study programs of air–ice–ocean interactive processes in the seasonal ice zone of the Greenland and Barents Seas', in 'Microwave remote sensing of sea ice' (American Geophysical Union, Washington, DC., 1992, Geophysical Monograph No. 68), pp. 261–89
68 JOHANNESSEN, O. M., SANDVEN, S., DROTTNING, A., KLOSTER, K., HAMRE, T., and MILES, M.: 'ERS-1 SAR sea ice catalogue' (European Space Agency, SP-1193, 1997)
69 KELL, P. E.: 'About bistatic RCS evaluation using results of monostatic RCS measurements', *Proceedings of IEEE*, 1965, **53** (8), pp. 1126–32
70 KELLER, J. B.: 'Geometrical theory of diffraction', *Journal of Optical Society of the America*, 1962, **52** (2), pp. 116–30
71 KOCK, W. E.: 'Pulse compression with periodic gratings and zone plane gratings', *Proceedings of IEEE*, 1970, **58** (9), pp. 1395–96
72 KONDRATENKOV, G. S.: 'The signal function of a holographic radar', *Radiotekhnika*, 1974, **29** (6), pp. 90–92 (in Russian)

73 KONDRATENKOV, G. S.: 'Synthetic aperture antennas', in VOSKRESENSKY, D. I. (Ed.): 'Phased antenna arrays design' (Radiotekhnika, Moscow, 2003), pp. 399–416 (in Russian)
74 KONDRATENKOV, G. S., POTEKHIN, V. A., REUTOV, A. P., and FEOKTISTOV, Yu. A.: 'Earth surveying radars' (Radio i Svyaz, Moscow, 1983) (in Russian)
75 KORSBAKKEN, E., JOHANNESSEN, J. A., and JOHANNESSEN, O. M.: 'Coastal wind field retrievals from ERS synthetic aperture radar images', *Journal of Geophysical Research*, 1998, **103** (C4), pp. 7857–74
76 KORSNES, R.: 'Some concepts for precise estimation of deformations/rigid areas in polar pack ice based on time series of ERS-1 SAR images', *International Journal of Remote Sensing*, 1994, **15** (18), pp. 3663–74
77 KRAMER, H.: 'Observation of the Earth and its Environment. Survey of Missions and Sensors' (Springer, Berlin, 1996)
78 KURIKSHA, A. A.: 'Moving target 2D radar imaging by combination of the aperture synthesis and tomography', *Radiotekhnika i Electronica*, 1994, **39** (4), pp. 613–18 (in Russian)
79 KWOK, R., and CUNNINGHAM, G. F.: 'Seasonal ice area and volume production of the Arctic Ocean: November 1996 through April 1997', *Journal of Geophysical Research*, 2002, **107** (C10), pp. 8038–42
80 LANDSBERG, G. S.: 'Optics' (Nauka, Moscow, 1970, 6th edn) (in Russian)
81 LARSON, R. W., ZELENKA, I. S., and IOHANSEN, E. L.: 'A microwave hologram radar system', *IEEE Transactions on Aerospace and Electronic Systems*, 1972, **AES-8** (2), pp. 208–17
82 LARSON, R. W., ZELENKA, I. S., and IOHANSEN, E. L.: 'Microwave holography', *Proceedings of IEEE*, 1969, **57** (12), pp. 2162–64
83 LARUE, A., HOFFMAN, K. N., HURLBUT, D. E., KIND, H. J., and WINTROUB A.: '94-GHz radar for space object identification', *IEEE Transactions on Microwave Theory and Techniques*, 1969, **MTT-17** (12), pp. 1145–49
84 LE HEGARAT-MUSCLE, S., ZRIBI, M., ALEM, F., WEISSE, A., and LOUMAGNE, C.: 'Soil moisture estimation from ERS/SAR data: Toward an operational methodology', *IEEE Transactions on Geoscience and Remote Sensing*, 2002, **GE-40** (12), pp. 2647–58
85 LEITH, E. N.: 'Quasi-holographic techniques in the microwave region', *Proceedings of IEEE*, 1971, **59** (9), pp. 1305–18
86 LEITH, E. N., and INGALLS, F. L.: 'Synthetic antenna data processing by wavefront reconstruction', *Applied Optics*, 1968, **7** (3), pp. 539–44
87 LEITH, E. N.: 'Side-looking synthetic aperture radar', in CASASENT, D. (Ed.): 'Optical data processing applications' (Springer-Verlag, Berlin, Heidelberg, New York, 1978) Chapter 4
88 LEWITT, P. M.: 'Reconstruction algorithms: transform methods', *Proceedings of IEEE*, 1983, **71** (3), pp. 390–408
89 LIKHACHEV, V. P., and PASMUROV, A. Ya.: 'Aircraft radar imaging under signal partial coherence conditions', *Radiotekhnika i Electronica*, 1999, **44** (3), pp. 294–300 (in Russian)

90 MAYZELS, E. N., and TORGOVANOV, V. A.: 'Measurement of scattering characteristics of radar targets' (Sovetskoe Radio, Moscow, 1972) (in Russian)
91 MEHRHOLZ, D., and MAGURA, K.: 'Radar tracking and observation of non-cooperative space objects by reentry of Salut-7-Kosmos-1686'. Proceedings of International workshop of European Space Operations Center, Darmstadt, Germany, 1991, pp. 1–8
92 MEIER, R. W.: 'Magnification and aberration three order of diffraction in holography' *Journal of the Optical Society of America*, 1965, **55** (7), pp. 987–91
93 MELLING, H.: 'Detection of features in first-year pack ice by synthetic aperture radar (SAR)', *International Journal of Remote Sensing*, 1998, **19** (6), pp. 1223–49
94 MENSA, D. L.: 'High resolution radar cross-section imaging' (Artech House, Dedham, USA, 1991)
95 MERSEREA, R. M., and OPPENHEIM, A. V.: 'Digital reconstruction of multidimensional signals from their projections', *Proceedings of IEEE*, 1974, **62** (10), pp. 1319–38
96 MILER, M.: 'Holography' (SNTL, Prague, Czechoslovakia, 1974) (in Czech)
97 MILES, V. V., BOBYLEV, L. P., MAKSIMOV, S. V., JOHANNESSEN, O. M., and PITULKO, P. M.: 'An approach for assessing boreal forest conditions based on combined use of satellite SAR and multispectral data', *International Journal of Remote Sensing*, 2003, **24** (22), pp. 4447–66
98 MITSMAKHER, M. Yu., and TORGOVANOV, V. A.: 'Microwave anechoic chambers' (Radio i Svyaz, Moscow, 1982) (in Russian)
99 MOORE, R. K.: 'Tradeoff between picture element dimensions and noncoherent overaging in side-looking airborne radar', *IEEE Transactions on Aerospace and Electronic Systems,* 1979, **AES-15** (5), pp. 697–708
100 MUNSON, D. C., JR., O'BRIEN, J. D., and IENKINS, W. K.: 'A tomographic formulation of spotlight-mode synthetic aperture radar', *Proceedings of IEEE*, 1983, **71** (8), pp. 917–25
101 NGHIEM, S.: 'On the use of ENVISAT ASAR for remote sensing of sea ice'. Abstracts of ENVISAT Symposium, Salzburg, Austria, 2004, Abstract No. 672
102 ONSTOTT, R. G.: 'SAR and scatterometer signatures of sea ice', in CARSEY, F. (Ed.): 'Microwave remote sensing of sea ice' (AGU Geophysical Monograph 68, AGU, 1992), pp. 73–104
103 PAPOULIS, A.: 'Systems and transforms with applications in optics' (McGraw-Hill, New York, 1968)
104 PASMUROV, A. Ya.: 'Aircraft radar imaging', *Zarubezhnaya Radioelectronica*, 1987, **12**, pp. 3–30 (in Russian)
105 PASMUROV, A. Ya.: 'Microwave holographic process modelling based on the edge waves method', *Radiotehnika i Electronica*, 1971, **26** (10), pp. 2030–33 (in Russian)
106 PASMUROV, A. Ya.: 'Tomographic methods for radar imaging'. Proceedings of the first All-Union conference on Optical information processing, Leningrad, USSR, 1988, pp. 85–86 (in Russian)

107 PETTERSSON, L. H., SANDVEN S., DALEN, O., MELENTYEV, V. V., and BABICH, N. G.: 'Satellite radar ice monitoring for ice navigation of the ARCDEV tanker convoy in the Kara sea'. Proceedings of the fifteenth international conference on Port and ocean engineering under Arctic conditions, Espoo, Finland, 1999, vol. 1, pp. 181–90
108 POLYANSKY, V. K., and KOVALSKY, L. V.: 'Information content of optical radiation'. Proceedings of the third III All-Union School on Holography, Leningrad, USSR, 1972, pp. 53–71 (in Russian)
109 POPOV, S. A., ROZANOV, B. A., ZINOVIEV, J. S., and PASMUROV, A. Ya.: 'Basic principles of microwave holograms inverse synthesis'. Proceedings of the eighth III All-Union School on Holography, Leningrad, USSR, 1976, pp. 275–89 (in Russian)
110 PORCELLO, L. J.: 'Turbulence-induced phase errors in synthetic aperture radars', *IEEE Transactions on Aerospace and Electronic Systems*, 1970, **AES-6** (5), pp. 634–44
111 RAMSAY, B. R., WEIR, L., WILSON, K., and ARKETT, M.: 'Early results of the use of RADARSAT ScanSAR data in the Canadian Ice Service'. Proceedings of the fourth Symposium on Remote sensing of the polar environments, Lyngby, Denmark, 1996, ESA SP-391, pp. 95–117
112 RANEY, R. K.: 'SAR processing of partially coherent phenomena', *International Journal of Remote Sensing*, 1980, **1** (1), pp. 29–51
113 RINO, C. L., and FREMOUW, E. J.: 'The angle dependence of singly scattered wave fields', *Journal of Atmospheric and Terrestrial Physics*, 1977, **39** (5), pp. 859–68
114 RINO, C. L., and OWEN, J.: 'Numerical simulations of intensity scintillation using the power low phase screen model', *Radio Science*, 1984, **19** (3), pp. 891–908
115 RINO, C. L., GONZALEZ, V. H., and HESSING, A. R.: 'Coherence bandwidth loss in transionospheric radio propagation', *Radio Science*, 1981, **16** (2), pp. 245–55
116 RINO, C. L.: 'On the application of phase screen models to the interpretation of ionospheric scintillation data', *Radio Science*, 1982, **17** (4), pp. 855–67
117 ROBINSON, I. S.: 'Measuring the oceans from space. The principles and methods of satellite oceanography' (Springer-Praxis, Chichester, UK, 2004)
118 RULE, M.: 'Radio telescopes of large resolving power', *Review of Modern Physics*, 1975, **47** (7), pp. 557–66
119 SANDVEN, S., DALEN, O., LUNDHAUG, M., KLOSTER, K., ALEXANDROV, V. Y., and ZAITSEV, L. V.: 'Sea ice investigations in the Laptev sea area in late summer using SAR data', *Canadian Journal of Remote Sensing*, 2001, **27** (5), pp. 502–16
120 SANDVEN, S., JOHANNESSEN, O. M., MILES, M. W., PETTERSSON, L. H., and KLOSTER, K.; 'Barents sea seasonal ice zone features and processes from ERS-1 synthetic aperture radar: Seasonal ice zone experiment 1992', *Journal of Geophysical Research*, 1999, **104** (C7), pp. 15843–57

121 SAPHRONOV, G. S., and SAPHRONOVA, A. P.: 'An introduction to microwave holography' (Sovetskoe Radio, Moscow, 1973) (in Russian)
122 SCHEUCHL, B., CAVES, R., FLETT, D., DE ABREU, R., ARKETT, M., and CUMMING, I.: 'The potential of cross-polarization information for operational sea ice monitoring'. Abstracts of ENVISAT Symposium, Salzburg, Austria, 2004, Abstract No. 493
123 SEA ICE INFORMATION SERVICES IN THE WORLD. WMO N 574. Secretariat of the World Meteorological Organization, Geneva, Switzerland, 2000
124 SEKISTOV, V. N., GAVRIN, A. L., ANDREEV, V. Yu. *et al.*: 'Low-orbit satellite radar imaging with narrow-band signals', *Radiotehnika i Electronica*, 2000, **45** (7), pp. 830–36 (in Russian)
125 SEPHTON, A. J., and PARTINGTON, K. C.: 'Towards operational monitoring of Arctic sea ice by SAR', in TSATSOULIS, C., and KWOK, R. (Eds): 'Analysis of SAR data of the polar oceans. Recent advances' (Springer-Praxis, Berlin, Heidelberg, 1998), pp. 259–79
126 SHUCHMAN, R. A., LYZENGA, D. R., LAKE, B. M., HUGHES, B. A., GASPAROVICH, R. F., and KASISCHKE, E. S.: 'Comparison of joint Canada–U.S. ocean wave investigation project syntetic aperture radar data with internal wave observations and modeling results', *Journal of Geophysical Research*, 1988, **93** (C10), pp. 12283–91
127 SKADER, G. D.: 'An introduction to computerized tomography', *Proceedings of IEEE*, 1978, **66** (6), pp. 5–16
128 SOH, L.-K., TSATSOULIS, C., and HOLT, B.: 'Identifying ice floes and computing ice floe distribution in SAR images', in TSATSOULIS, C., and KWOK, R. (Eds): 'Analysis of SAR data of the polar oceans. Recent advances' (Springer-Praxis, Berlin, Heidelberg, 1998), pp. 9–34
129 STEINBERG, B. D.: 'Microwave imaging with large antenna arrays. Radio camera principles and techniques' (John Wiley & Sons, New York, 1983)
130 STEINBERG, B. D.: 'Aircraft radar imaging with microwaves', *Proceedings of IEEE*, 1988, **76** (12), pp. 1578–92
131 STROKE, G. W.: 'An introduction to coherent optics and holography' (Academic Press, New York, London, 1966)
132 TATARSKY, V. I.: 'Wave propagation in a turbulent atmosphere' (Nauka, Moscow, 1967) (in Russian)
133 TATARSKY, V. I.: 'Wave propagation in a turbulent medium' (McGraw-Hill, New York, 1961)
134 THOMPSON, M. C., and JANES, H. B.: 'Measurements of phase front distortion on an elevated line-of-sight path', *IEEE Transactions on Aerospace and Electronic Systems*, 1970, **AES-6** (5), pp. 645–56
135 TISON, C., NICOLAS, J.-M., TUPIN, F., and MAITRE, H.: 'A new statistical model for Markovian classification of urban areas in high-resolution SAR images', *IEEE Transactions on Geoscience and Remote Sensing*, 2004, **GE-42** (10), pp. 2046–57

136 TITOV, M. P., TOLSTOV, E. F., and FOMKIN, B. A.: 'Mathematical modelling in aviation', in BELOCERKOVSKY, S. M. (Ed.): 'Problems of cybernetics' (Nauka, Moscow, 1983), pp. 139–45

137 UFIMTZEV, P. Ya.: 'Method of edge waves in physical diffraction theory' (Sovetskoe Radio, Moscow, 1962) (in Russian)

138 VARGANOV, M. E., ZINOVIEV, J. S., ASTANIN, L. Yu. *et al.*: 'Aircraft radar characteristics' (Radio i Svyaz, Moscow, 1985) (in Russian)

139 WIRTH, W. D.: 'High resolution in azimuth for radar targets moving on a straight line', *IEEE Transactions on Aerospace and Electronic Systems*, 1980, **AES-16** (1), pp. 101–3

140 WALKER, I. L.: 'Range-Doppler imaging of rotating objects', *IEEE Transactions on Aerospace and Electronic Systems*, 1980, **AES-16** (1), pp. 23–52

141 YEH, K. C., and LIN, C. H.: 'Radio wave scintillation in the ionosphere', *Proceedings of IEEE*, 1982, **70** (4), pp. 324–60

142 YU, F. T. S.: 'Introduction to diffraction, information processing, and holography' (The MIT Press, Cambridge, MA, 1973)

143 ZINOVIEV, J. S., and PASMUROV, A. Ya.: 'Holographic principles application for SAR analysis', in POTEKHIN, V. A. (Ed.): 'Image and signal optical processing' (USSR Academy of Sciences, Leningrad, 1981), pp. 3–15 (in Russian)

144 ZINOVIEV, J. S., and PASMUROV, A. Ya.: 'Evaluation of SAR phase fluctuations caused by turbulent troposphere', *Radiotehnika i Electronica*, 1975, **20** (11), pp. 2386–88 (in Russian)

145 ZINOVIEV, J. S., and PASMUROV, A. Ya.: 'Method for recording and processing of 1D Fourier microwave holograms', *Pisma v Zhurnal Tekhnicheskoy Fiziki*, 1977, **3** (1), pp. 28–32 (in Russian)

146 ZINOVIEV, J. S., and PASMUROV, A. Ya.: 'Methods of inverse aperture synthesis for radar with narrow-band signals', *Zarubezhnaya Radioelectronica*, 1985, **3**, pp. 27–39 (in Russian)

List of abbreviations

1D	One-dimensional
2D	Two-dimensional
3D	Three-dimensional
AB	Adaptive beamforming
AEC	Anechoic chamber
CAT	Computer-aided tomography
CBP	Convolutional backprojection method
CCA	Circular convolution algorithm
CIS	Canadian Ice Centre
DFT	Discrete Fourier transform
ECP	Extended coherent processing
ESA	European Space Agencies
EWM	Edge waves method
FCC	Frequency contrast characteristics
FFT	Fast Fourier transform
GSSR	Goldstone solar system radar
GTD	Geometrical theory of diffraction
IFT	Inverse Fourier transform
ISAR	Inverse synthetic aperture radar
LFM	Linear frequency modulation
LRIR	Long-range imaging radar
NRCS	Normalised radar cross-section
NBM	Narrowband mode
PH	Partial hologram
PRR	Pulse repetition rate
RCS	Radar cross-section
RLOS	Radar line of sight
SAP	Synthetic antenna pattern
SAR	Synthetic aperture radar
SCF	Space carrier frequency
SCS	Specific cross-section
SGL	Spatial grey level
SST	Sea surface temperature
WBM	Wideband mode
WMO	World Meterological Organization

Index

Abbe's formula 21, 108
adaptive beamforming 33
adaptive beamforming algorithm 21
aerodynamic target 148, 151–2
airborne radars 60, 79, 196
aircraft 60
aircraft imaging 21–3, 215
algorithms
   adaptive beamforming 21
   Calman's 186
   circular convolution 34–5
   convolution back-projection 18–19, 73–5, 118–19
   heuristic 187–8
   interpolation 18, 73–5
   processing 130–45
   range-Doppler 34–5
   reconstruction 221
   tomographic 70, 72–7
   Wiener 185–6
all-weather mapping 24
Almaz-1 192, 195
amplitude factor 12
anechoic camera 114, 116
anechoic chamber 30, 113, 218–22
   echo-free zone 218
   reconstruction algorithm 221
Antarctic 195
antenna approach 33, 147
antenna arrays 20–2, 60
antirecognition devices 229
aperture angle 33
aperture characteristics 173–8
aperture noise 176
aperture performance 173

aperture synthesis 31–3
aposterior techniques 184–5
archaeological surveys 24
Arctic 195–206
   sea ice monitoring 195–8
artificial reference wave 38–9, 51
ASAR 193–4, 196–7, 205
   operation modes 193–4
aspect variation 126
autocorrelation coherence 84
autocorrelation function 84
averaging of resolution elements 87, 94
azimuth ambiguity function 172
azimuth defocusing 205–6
azimuthal resolution 180–1
azimuth-range 49

back projection 131
bathymetry 195
Bayes classifier 226, 228
Bessel functions 29
bistatic radar 101–2
bistatic scattering 28–9

Calman's filtering algorithms 186
Carman's model 160–2
carrier track instabilities 57–8
CAT radar 76
circular convolution algorithm 34–5
classification: *see* target classification
cloud effects 166
coherence 40
coherence length 40
coherence stability 40–1, 43
coherent imaging 47–8

coherent radar 21–3
  holographic processing 36–41
  tomographic processing 41–8
coherent signal 40–1
coherent summation of partial components 126–31
  1D 139
  2D viewing geometry 131–42
  3D viewing geometry 141–5
  complexity 136–7, 140–2, 145
complex microwave Fourier hologram 110–15
complex targets 27–8
computer-aided tomography 74, 76
computerised tomography 14–20, 48
  remote-probing 15
  remote-sensing 15
  see also tomographic processing
contrast 94–9, 175, 177
convolution back-projection algorithm 18–19, 73–5, 118–19
correlated processing 35, 49
correlation function 96
critical volume 179
cross range resolution 148–51
cross-correlation approach 33–4
cylinder 29–31, 219, 221–4
  local scattering characteristics 223–4

dark level 175, 177
deformed ice 201, 204
density distribution 14–16, 19
diffraction 29, 116
diffraction-limited image 127
digital processing 112–16, 145
direct synthesis 31–2
distortion 176
Doppler frequency shift 27
Doppler-range method: see range-Doppler method
dynamic range 175, 177

earth surface imaging 20, 34, 60
  satellite SARs 191–215
earth surface survey 34, 70–1, 79
echo signal 27, 46, 148, 182–3
edge wave method 29
electron density fluctuations 166–7
ENVISAT 193–4, 196–7, 205
ERS-1 20, 193, 195, 197, 212
ERS-2 20, 193, 195, 197, 206, 208, 210–11, 214
  mesoscale ocean phenomena 208, 210–11, 214
  sea ice 206
extended coherent processing 35–6
extended targets 28–9, 31, 79–85
  compact 28–9
  partially coherent 85–6
  proper 28, 31

fast ice 201, 204
first-year ice 200–2
flop 136
focal depth 8–10, 14, 67–70
focal length 7
focal point 7
focused aperture 54
focusing depth 59
forestry 195
Fourier microwave hologram 39–40, 52–3
  complex 110–15
  rotating target 101–9
  simulation 112–16
Fourier space 16, 18
Fourier transform 18
Fraunhofer microwave hologram 39–40, 52–3
frequency stability 40–1
frequency-contrast characteristic 95–8
Fresnel lens 49
Fresnel microwave hologram 39–40, 52
Fresnel zone plate 33, 50
Fresnel-Kirchhoff diffraction formula 38
friction velocity 207
front-looking holographic radar 60–70
  hologram recording 60–3
  image reconstruction 62–7
  resolution 61–2

gain in the signal-to-noise ratio 174–5
geological structures 24
geometric accuracy 80
geometrical theory of diffraction 29
globules 158–9
Goldstone Solar System Radar 41
grease ice 198–9
grey-level resolution 178–81

half-tone resolution 178–81
Hankel transform 75–6

Index

heuristic algorithm 187–8
hologram 11–14
  real image 12–13, 49–50, 62–6
  virtual image 12–13, 49–50, 62–6
  wideband 123–4
  see also microwave hologram
hologram function 11–12
hologram modulation index 12
hologram recording 10–11
  1D 51
  front-looking holographic radar 60–3
  SAR 50–3
holographic image 14
holographic processing
  coherent radar 36–41
  front-looking radar 60–3
  ISAR 35–6
  rotating targets 101–16
  SAR 33–4
holographic technique 1–2
holography 10–14
homomorphic image processing 186
Huygens-Fresnel integral 53

ice edge 201–2, 205
ice floes 200–2
ice monitoring 195–8
ice navigation 196–7, 201
ice parameters 195, 197
icebergs 195–6, 202, 206
icebreakers 197–8, 201
ICEWATCH 196
image
  computerised tomography 14–20
  holographic 10–14
  microwave 20–5
  optical 7–10
  thin lens 7–8
image intensity 81, 87–96
image interpretability 178–80
image interpretation 80
image quality 24, 57–60, 77, 80–2, 173–81
  integral evaluation 177–81
image reconstruction 11–14, 16, 36, 124
  coherent summation of partial components 126–30
  digital simulation 112–16
  front-looking holographic radar 62–7
  microwave hologram 53–6
  spotlight SAR 72–7

image smoothing 88, 91, 93–4
image stability 174
imaging radars 7
imaging time 176
impulse response 152–3
incoherent signal integration 81, 87, 90–4
inertia region 159–61
INMARSAT 196
integral image 132–5
interference pattern 11, 13
internal waves 212–14
interpolation algorithm 18, 73–5
interpretability 178–80
intrinsic aperture noise level 176
inverse aperture synthesis 23, 147–8, 215–17
inverse Fourier transform 118
inverse source problem 16
inverse synthesis 31–2
  rotating target 101–9
inverse synthetic aperture radar: *see* ISAR
ionosphere 147
  electron density fluctuations 166–7
  turbulence 166–7, 172
  turbulence parameter 167
ISAR 32–3, 148
  instability 40–1
  signal processing 34–6
  tomographic processing 41

Kell's theorem 102
kernel function 139
Kosmos-1870 192, 195

linear filtering model 124
linear filtration theory 95, 98
linearly moving target 147
local responses 30
local statistics technique 186–8
long-range imaging radar 215–16
low contrast targets 94–9

magnification 8, 63–7
mean image power 175
median filtering 184
microholograms 13–14
micronavigation noise 173
microwave holograms 2, 36–40
  1D 101–12
  amplitude-phase 38
  Fourier 39–40, 46, 52–3, 101–16

microwave holograms (*continued*)
 Fraunhofer 39–40, 52–3
 Fresnel 39–40, 52
 multiplicative 37, 50
 narrowband 131–2, 134, 136–7, 140, 145
 phase-only 37–8
 quadrature 37–8, 106, 115
 wideband 133, 136–42
microwave holographic receiver 38–9
microwave image 7, 20, 23–4
microwave imaging 20–5, 101
microwave radars 7
microwaves 1
monostatic scattering 28–9
moving targets 58–9
 rotating 101–45
 straight line 147–56
multibeam processing: *see* multi-ray processing
multiplicative noise 98
multi-ray processing 87, 94, 184

narrowband microwave hologram 131–2, 134, 136–7, 140, 145
Newton's formulae 7–8
nilas 198, 200–1
noise dark level 175
noise distribution 177–8
nonparametric classifier 226, 228
normalised radar cross-section 205–7
Northern Sea Route 196–8

ocean circulation 194–5
ocean currents 94, 212
ocean dynamics 191
ocean phenomena
 mesoscale 204–15
 surface velocity 205
 see also sea surface imaging
ocean waves 191, 194–5
 internal 212–14
 see also wave imaging
oceanography 191–2
oil spills 94, 195, 211–12
old ice 200, 202
optical image 7–10, 23
 real 10
 virtual 10
orthoscopic image 10, 12, 14

pancake ice 200, 203
panoramic radars 20
partial coherence 79, 84
partial holograms 127–45
 spectral components 136
partial images 130, 134, 137, 140–3, 145
 radial 142
 transverse 137, 140–3
partially coherent signal 40, 148–51
 radar image modelling 152–6
partially coherent target imaging 79
 extended 83–6
 low contrast 94–9
 mathematical model 85–6
 statistical image characteristics 87–94
path instabilities 151–6
pattern recognition theory 225
phase 12, 14
phase errors 157–72
 turbulent ionosphere 172
 turbulent troposphere 167–72
phase fluctuations 167–70, 172
phase noise 152, 156
phase-only hologram 37–8
pixels 176
planet surveys 41, 217
plate contrast coefficient 12
point targets 27, 58
polar format processing 35–6
polar grid 18
potential functions 226, 228
potential SAR characteristics 173–5
principal planes 7–8
probing 14–18
projection slice theorem 17–18, 47, 72, 74
pseudoscopic image 10, 12, 14, 64

quasi-holographic radar systems 2, 31, 49–60
 hologram recording 50–3
 image reconstruction 53–6

radar characteristics 175–8
radar cross-section 28, 30
radar data processing 124–6
radar imaging 2
 basic concepts 7–25
 methods 27–48
 microwave 20–5
 partially coherent signals 152–6

Index   247

radar interferometer 217
radar responses 217–19
   closed tests 218–19
   open tests 218
RADARSAT 193–7, 201–2, 204
radio camera 21
radio telescope 16
radio vision 1
radiometric precision 80
radiometric resolution 81, 94
rain cells 212–13
range resolution 180–1
range-Doppler algorithm 34–5
range-Doppler method 1–3, 33, 215
Rayleigh model 182–3
real antennas 20
real apertures 20–1
recognition: *see* target recognition
reconstruction algorithm 221
reference voltage 22
reference wave 11–13
refractive index, troposphere 157–66
resolution 23, 33, 59–60, 177
   azimuthal 180–1
   cross range 148–51
   defocused microwave image 108–9
   front-looking holographic radar 61–2
   grey-level 178–81
   half-tone 178–81
   path instabilities 153–6
   potential 173
   radiometric 81, 94
   range 180–1
   spatial 80, 94
   spotlight SAR 76
   synthesised Fourier hologram 107–8
resolving power: *see* resolution
Rice reflection model 182–3
rotating target imaging
   holographic approach 101–16
   tomographic approach 117–45

sample characteristic 174
sampling theorem 21
SAR 31–2, 85–6
   holographic approach 33–4
   instability 40
   low contrast targets 94, 97–8

potential characteristics 173–5
satellite SARs 191–5
signal processing 33–4
spaceborne 167–8
test ground 176
turbulence 167–8
see also side-looking synthetic aperture
   radars
see also spot-light SAR
satellite imaging 79, 120–4, 129, 131, 143, 215
   aspect variation 120–1
satellite SARs 191–5
scaling 64–67, 69
ScanSAR 195–7, 201–2, 204
scatterers 28–31
scattering matrix 217
sea currents 94, 212
sea ice 192–3
   classification 198–204
   imagery 198–206
   monitoring 195–8
   parameters 198, 203–4
sea surface imaging 79–80, 99, 204–5
   rough sea surface 82–5
see also ocean phenomena
see also wave imaging
sea surface temperature 207, 209
SEASAT 191–2, 195
sharpness 175
ship wakes 214–15
ship wrecks 211
side-looking radar 1–3, 21
side-looking synthetic aperture radars 31–2, 49–60, 79
   hologram recording 50–3
   image reconstruction 53–6
   resolution redistribution 180–1
see also SAR
sigma-filter 187–8
sign vectors 225–8
signal processing 33–6
signal-to-noise ratio 88–90, 92–4
   gain 174–5
SIR-A 191–2
SIR-B 191–2, 195
SIR-C 192
Smith-Wentraub formula 157
soil classification 24
soil moisture 195
space carrier frequency 12, 109

space frequency 44–8
Space Shuttle 191–2
spaceborne SAR 167–8
spacecraft identification 79, 126, 215–17
  2D 215–17
  see also satellite imaging
spatial resolution 80, 94
specific cross-section 80–1, 95–6
speckle 24–5, 80, 175, 177, 181–9
  statistical characteristics 182–4
  suppression 184–9
speckle field 14
spot-light SAR 32, 70–7
  image reconstruction 72–7
  resolution 76
squall lines 212
statistical image characteristics 87–94
structure function 159, 163–5
subsurface probing 24
subwater imaging 24
surface hologram 124
swath width 176
swell 208–10
synthesis range 97–8
synthetic antenna 21–2
synthetic antenna pattern 173
synthetic aperture length 23
synthetic aperture pattern 23, 33, 173–4
synthetic aperture radar imaging 19
synthetic aperture radars: see SAR
synthetic apertures 20–3
  see also aperture synthesis

target characteristics 217–24
target classification 222–8
target models 27–31
target recognition 222–9
  efficiency 226, 228
  mathematical model 225–6
  probability 226, 228
  sign vectors 225–8
target reflectivity 43–5
target viewing 42
targets
  aerodynamic 148, 151–2
  complex 27–8
  extended 28–9, 31, 79–82

low contrast 94–9
moving 58–9
moving in a straight line 147–56
partially coherent 79, 83–99
point 27, 58
rotating 101–45
three-dimensional images 10, 12, 14, 20, 25, 69, 127
three-dimensional viewing geometry 119–26
tomographic algorithms
  spot-light SAR 70, 72–7
tomographic processing
  coherent radar 41–8
  frequency domain 117–18
  ISAR 35–6, 41
  rotating targets 117–45
  SAR 33
  space domain 118–19
  spot-light SAR 70–7
  see also computerised tomography
tomographic techniques 2
tomography 14
transmittance 12
troposphere 157–72
  near-earth 159
  phase errors 167–72
  refractive index distribution 157–66
  turbulence 158–72
true image 14
turbulence 158–63, 165–72
  inner-scale size 159
  ionosphere 166–7, 172
  isotropic 158
  outer-scale size 158
  troposphere 158–72
turbulent flows 148
two-dimensional image 127
two-dimensional viewing geometry 41–8
  rotating targets 131–42

uncertainty function 61
unfocused aperture 54
unistatic radar 101–3
upwelling 207–9
urban area imaging 24–5
urban area monitoring 195

velocity bunching effect 209–10

wave imaging 82–5, 205–6, 208–9
  see also ocean waves
whirls 159–61, 166–7, 172
wideband hologram 123–4
wideband microwave hologram 133, 136–42
  processing algorithms 138–9
Wiener filtering algorithm 185–6
Wiener-Khinchin theorem 166
wind slicks 94

wind squall 212
wind stress 208

X-ray imaging 2
X-ray tomography 17–19

young ice 198, 200–1